Information Systems Engineering and Management

Volume 33

The book series "Information Systems Engineering and Management" (ISEM) publishes innovative and original works in the various areas of planning, development, implementation, and management of information systems and technologies by enterprises, citizens, and society for the improvement of the socio-economic environment.

The series is multidisciplinary, focusing on technological, organizational, and social domains of information systems engineering and management. Manuscripts published in this book series focus on relevant problems and research in the planning, analysis, design, implementation, exploration, and management of all types of information systems and technologies. The series contains monographs, lecture notes, edited volumes, pedagogical and technical books as well as proceedings volumes.

Some topics/keywords to be considered in the ISEM book series are, but not limited to: Information Systems Planning; Information Systems Development; Exploration of Information Systems; Management of Information Systems; Blockchain Technology; Cloud Computing; Artificial Intelligence (AI) and Machine Learning; Big Data Analytics; Multimedia Systems; Computer Networks, Mobility and Pervasive Systems; IT Security, Ethics and Privacy; Cybersecurity; Digital Platforms and Services; Requirements Engineering; Software Engineering; Process and Knowledge Engineering; Security and Privacy Engineering, Autonomous Robotics; Human-Computer Interaction; Marketing and Information; Tourism and Information; Finance and Value; Decisions and Risk; Innovation and Projects; Strategy and People.

Indexed by Google Scholar. All books published in the series are submitted for consideration in the Web of Science.

For book or proceedings proposals please contact Alvaro Rocha (amrrocha@gmail.com).

Sandeep Kautish · Álvaro Rocha · Ankur Gupta ·
Sahil Sawhney
Editors

Strategy Analytics for Business Resilience Theories and Practices

Springer

Editors
Sandeep Kautish
Apex Institute of Technology (AIT CSE)
Chandigarh University
Chandigarh, Punjab, India

Álvaro Rocha
Information Systems
University of Lisbon
Lisboa, Portugal

Ankur Gupta
Model Institute of Engineering
and Technology (MIET)
Jammu, Jammu and Kashmir, India

Sahil Sawhney
Strategic Initiatives
Model Institute of Engineering
and Technology (MIET)
Jammu, Jammu and Kashmir, India

ISSN 3004-958X ISSN 3004-9598 (electronic)
Information Systems Engineering and Management
ISBN 978-3-031-82368-8 ISBN 978-3-031-82369-5 (eBook)
https://doi.org/10.1007/978-3-031-82369-5

This Springer imprint is published by the registered company Springer Nature Switzerland AG
The registered company address is: Gewerbestrasse 11, 6330 Cham, Switzerland

Contents

Enhancing Organizational Success: A Strategic Approach to Data Quality and Governance

Pawan Whig, Nikitha Yathiraju, Anupriya Jain, Alpna Sharma, and Sandeep Kautish

Abstract In the era of big data and advanced analytics, the quality and governance of data have become pivotal for organizational success. This chapter explores the strategic importance of data quality and governance, highlighting their roles in ensuring accurate, complete, and reliable data for informed decision-making. It delves into the key components of a robust data governance framework, including data stewardship, data quality management, and regulatory compliance. Through case studies and best practices, the chapter provides actionable insights for organizations aiming to enhance their data governance strategies. By addressing common challenges and presenting solutions, this chapter serves as a comprehensive guide for leaders and data professionals striving to maximize the value of their data assets.

Keywords Data quality · Data governance · Data stewardship · Data management · Regulatory compliance · Big data · Advanced analytics · Decision-making · Data framework · Best practices · Data challenges · Organizational success

---

P. Whig (✉) · A. Sharma
Vivekananda Institute of Professional Studies, Technical Campus, Delhi, India
e-mail: pawanwhig@gmail.com

N. Yathiraju
University of the Cumberland, Williamsburg, USA

A. Jain
SCA, MRIIRS, Faridabad, India

S. Kautish
Chandigarh University, Mohali, India

S. Kautish et al. (eds.), *Strategy Analytics for Business Resilience Theories and Practices*, Information Systems Engineering and Management 33,
https://doi.org/10.1007/978-3-031-82369-5_1

1 Introduction

In today's data-driven world, organizations are inundated with vast amounts of information generated from various sources [1]. This data, if managed and utilized effectively, can offer profound insights that drive strategic decisions, enhance operational efficiency, and foster innovation. However, the potential benefits of data are heavily contingent upon its quality and governance. Without proper mechanisms to ensure data accuracy, completeness, and reliability, organizations risk making decisions based on flawed information, which can lead to suboptimal outcomes or even significant strategic missteps. Therefore, data quality and governance have emerged as crucial pillars for any organization aiming to harness the power of data [2–5].

Data quality refers to the condition of data based on factors such as accuracy, completeness, reliability, and relevance. High-quality data is essential for several reasons:

1. **Informed Decision-Making**: Decisions based on high-quality data are more likely to be accurate and effective. Poor data quality can lead to erroneous conclusions and actions, potentially causing financial losses, reputational damage, and missed opportunities.
2. **Operational Efficiency**: High-quality data streamlines operations by reducing errors and rework. This leads to more efficient processes and cost savings.
3. **Customer Satisfaction**: Accurate and reliable data helps organizations better understand and meet customer needs, thereby enhancing customer satisfaction and loyalty.
4. **Compliance and Risk Management**: Regulatory requirements often mandate accurate and complete data records. High-quality data ensures compliance with such regulations, thereby mitigating legal and financial risks.

Data governance involves the management of data availability, usability, integrity, and security within an organization [6–8]. It encompasses policies, procedures, and standards that ensure data is managed and used effectively and ethically.

Integrating data quality and governance into the organizational strategy is essential for several reasons:

1. **Competitive Advantage**: Organizations that effectively manage data quality and governance can gain a competitive edge by leveraging accurate insights to drive innovation and strategic initiatives.
2. **Trust and Credibility**: High-quality data and robust governance practices build trust among stakeholders, including customers, partners, and regulators. This trust is crucial for maintaining a positive reputation and fostering long-term relationships.
3. **Agility and Adaptability**: A well-governed data environment enables organizations to respond quickly to market changes and emerging opportunities. Reliable data supports agile decision-making and strategic pivots.
4. **Resource Optimization**: Efficient data management practices reduce waste and optimize the use of resources, leading to cost savings and better allocation of

investments. This proactive approach minimizes potential damages and enhances organizational resilience.

Despite the clear benefits, many organizations struggle with data quality and governance. Common challenges include:

1. **Data Silos**: Disparate systems and departments often lead to fragmented data, making it difficult to ensure consistency and accuracy.
2. **Cultural Resistance**: Implementing data governance requires a cultural shift towards recognizing data as a strategic asset. Resistance from employees can hinder these efforts.
3. **Resource Constraints**: Effective data governance requires investment in technology, personnel, and processes. Limited resources can impede the development and implementation of robust governance frameworks.
4. **Complexity of Regulations**: Navigating the complex landscape of data regulations can be challenging, especially for organizations operating in multiple jurisdictions.
5. **Rapid Technological Changes**: The fast pace of technological advancements necessitates continuous adaptation of data governance practices to keep up with new tools and methodologies.

Data quality and governance are indispensable for organizations seeking to thrive in a data-centric world [9, 10]. By ensuring that data is accurate, reliable, and well-managed, organizations can make informed decisions, optimize operations, and maintain regulatory compliance. The strategic integration of data quality and governance not only mitigates risks but also unlocks opportunities for innovation and growth. As data continues to grow in volume and complexity, the importance of robust data governance frameworks will only increase, making it imperative for organizations to prioritize these aspects in their strategic planning.

2 Understanding Data Quality

Data quality refers to the condition of data based on various factors that determine its fitness for use. Key concepts include:

1. **Accuracy**: The extent to which data correctly describes the real-world object or event it is intended to represent. Accurate data is free from errors and discrepancies.
2. **Completeness**: The degree to which all required data is present. Incomplete data can lead to incorrect conclusions and poor decision-making.
3. **Consistency**: Ensures that data is uniform across different databases and systems, preventing discrepancies in information.
4. **Reliability**: Refers to the dependability of the data over time. Reliable data produces consistent results under consistent conditions.

5. **Validity**: Ensures that data values fall within the expected ranges and meet the required formats and standards.

3 Dimensions of Data Quality

Understanding the different dimensions of data quality helps in assessing and improving it. The main dimensions include:

1. **Accuracy**: The correctness of data. For example, a customer's address should be accurate to ensure successful deliveries.
2. **Completeness**: All necessary data should be present. Missing data, such as a missing phone number in a contact list, can affect operations.
3. **Consistency**: Data should be the same across different datasets. For instance, a customer's name should be spelled the same way in all databases.
4. **Timeliness**: Data should be up-to-date. For example, stock levels should be current to prevent over-ordering or stockouts.
5. **Uniqueness**: There should be no duplicate records. Duplicate customer records can lead to redundant communication and inefficiencies.
6. **Validity**: Data should conform to defined formats and constraints. For example, a date field should only contain valid dates.
7. **Integrity**: Data should be correctly linked. For example, a product ID should correctly correspond to the product details in another table.
8. **Relevance**: Data should be applicable to the context it is used in. Irrelevant data can mislead analyses and decisions.

4 Measuring Data Quality

To ensure high data quality, it must be measured accurately. Key methods include:

1. **Data Validation**: Applying rules and checks during data entry to ensure accuracy and consistency. For example, verifying email formats during entry.
2. **Key Performance Indicators (KPIs)**: Setting KPIs for data quality dimensions. For instance, measuring the percentage of duplicate records or missing values.
3. **Data Quality Assessment Tools**: Utilizing software tools that can automate the assessment of data quality across various dimensions and generate reports.
4. **Surveys and Feedback**: Collecting feedback from users regarding data quality issues they encounter. This helps in identifying and addressing user-specific problems.

5 Common Data Quality Issues

Several issues can affect data quality as shown in Fig. 1, including:

Fig. 1 Common data quality issues

1. **Data Entry Errors**: Mistakes made during data entry can result in inaccurate or inconsistent data.
2. **Duplicate Records**: Multiple entries for the same entity can lead to inefficiencies and confusion.
3. **Incomplete Data**: Missing information can hinder decision-making and operational processes.
4. **Outdated Data**: Stale data can lead to poor decisions based on incorrect assumptions.
5. **Inconsistent Data**: Variations in data across different systems can lead to confusion and errors.
6. **Data Silos**: Isolated data systems prevent comprehensive data analysis and can cause inconsistencies.
7. **Incorrect Data Mapping**: Poor integration between systems can result in data being misinterpreted or misplaced.
8. **Non-standardized Data**: Lack of standardized formats and definitions can make data difficult to interpret and use.

6 Tools and Techniques for Ensuring Data Quality

Maintaining high data quality involves using various tools and techniques, including and Data Quality Process is shown in Fig. 2

Fig. 2 Data quality process

1. **Data Cleaning Tools**: Software that can identify and correct errors, remove duplicates, and fill in missing values. Examples include OpenRefine, Talend, and Informatica.
2. **Data Integration Tools**: Tools that ensure data consistency across different systems by managing data flow and transformations. Examples include Apache Nifi, MuleSoft, and IBM DataStage.
3. **Data Quality Management Tools**: Comprehensive platforms that offer profiling, auditing, validation, and monitoring capabilities.
4. **Master Data Management (MDM)**: Systems that centralize and manage critical data to ensure consistency and accuracy across the organization. Examples include Oracle MDM, SAP Master Data Governance, and Informatica MDM.
5. **Data Governance Frameworks**: Establishing policies, roles, and responsibilities for data management to ensure compliance and accountability.
6. **Regular Audits and Monitoring**: Conducting periodic reviews of data quality and implementing continuous monitoring to identify and resolve issues promptly.
7. **Training and Education**: Ensuring that staff are aware of the importance of data quality and are trained in best practices for data entry and management.
8. **Automated Workflows**: Implementing automated processes to handle repetitive data quality tasks, such as data validation and cleansing, to reduce the risk of human error.

By understanding and addressing these aspects, organizations can significantly improve their data quality and, consequently, their ability to make informed and effective decisions [11–15].

7 Data Governance Framework

It involves establishing policies, procedures, and standards for how data is used, stored, and protected within an organization [16–19]. The primary goals of data governance are to ensure data quality, improve data security, and enhance the overall management of data as a valuable asset.

7.1 Roles and Responsibilities in Data Governance

Effective data governance requires clear roles and responsibilities:

1. **Data Governance Council**: Provides strategic direction, approves data policies, and ensures alignment with business goals. Members include senior executives and representatives from key departments.
2. **Chief Data Officer (CDO)**: Leads the data governance initiative, oversees the development and implementation of data governance policies, and ensures that data governance aligns with organizational objectives.
3. **Data Stewards**: Manage data assets within specific domains, ensuring data quality and compliance with governance policies. They act as liaisons between the data governance council and operational teams.
4. **Data Owners**: Individuals or departments responsible for specific datasets. They define data requirements, approve data access requests, and ensure data is used appropriately.
5. **Data Custodians**: IT personnel responsible for the technical aspects of data management, including data storage, security, and backups.
6. **Data Users**: Employees who access and use data for their daily tasks. They must adhere to data governance policies and report any data quality issues or security concerns.

7.2 Policies and Procedures

Establishing clear policies and procedures is crucial for effective data governance:

1. **Data Classification Policy**: Defines how data should be categorized based on sensitivity, value.

2. **Data Access Policy**: Specifies who can access data, under what conditions, and for what purposes. This includes user authentication, authorization, and auditing mechanisms.
3. **Data Retention Policy**: Defines how long data should be retained and the procedures for archiving and disposing of data. This ensures compliance with legal and regulatory requirements.
4. **Data Privacy Policy**: Outlines how personal data should be handled to protect individuals' privacy rights. This includes procedures for data anonymization, consent management, and data subject access requests.
5. **Data Usage Policy**: Defines acceptable and unacceptable uses of data within the organization. This helps prevent data misuse and ensures data is used ethically and legally.
6. **Incident Management Procedure**: Details the steps to be taken in case of a data breach or other data-related incidents. This includes notification, investigation, mitigation, and reporting procedures.

7.3 Implementing a Data Governance Framework

Implementing a data governance framework involves several steps:

1. **Assessment and Planning**: Conduct a thorough assessment of the current data management practices, identify gaps,. Develop a detailed implementation plan, including timelines, resources, and milestones.
2. **Stakeholder Engagement**: Engage stakeholders from across the organization to ensure their buy-in and support.
3. **Policy Development**: Develop and document data governance policies and procedures based on best practices and regulatory requirements. Ensure that policies are clear, comprehensive, and easily accessible.
4. **Framework Establishment**: Set up the data governance council, appoint data stewards. Establish processes for data quality management, data security, and data lifecycle management.
5. **Technology Implementation**: Ensure that these tools are integrated with existing systems and workflows.
6. **Training and Education**: Provide training and resources to employees to ensure they understand and comply with data governance policies. This includes training on data management best practices, security protocols, and regulatory requirements.
7. **Monitoring and Reporting**: Implement mechanisms to monitor data governance activities and measure their effectiveness. Regularly review and update policies and procedures to address emerging challenges and opportunities.
8. **Continuous Improvement**: Foster a culture of continuous improvement by regularly assessing the data governance framework, gathering feedback from stakeholders, and making necessary adjustments. This ensures that the framework remains effective and aligned with organizational goals.

By establishing a comprehensive data governance framework, organizations can ensure that their data is managed effectively, securely, and in compliance with regulations, thereby maximizing the value of their data assets.

8 Data Stewardship and Management

Effective data stewardship involves adhering to best practices that ensure data is managed efficiently and effectively [20–22]. One key practice is establishing clear roles and responsibilities for data stewards, ensuring they have the authority and resources needed to perform their duties. Regular training and education are essential to keep data stewards updated on the latest data management techniques and regulatory changes. Implementing robust data quality management processes, such as data profiling, cleansing, and validation, helps maintain high data quality standards. Additionally, fostering a culture of collaboration and communication between data stewards and other stakeholders ensures that data governance policies are understood and followed across the organization. By adopting these best practices, organizations can enhance their data stewardship efforts and improve overall data quality [23–25].

9 Data Management Principles

Data management principles provide a foundation for effective data governance and stewardship. These principles include ensuring data accuracy, completeness, consistency, and timeliness. Data should be collected, stored, and processed in a manner that maintains its integrity and reliability. Standardizing data formats and definitions across the organization helps ensure consistency and facilitates data integration. Data security is another critical principle, involving the protection of data from unauthorized access, breaches, and other security threats. Transparency and accountability are also important, with clear documentation of data management processes and regular audits to ensure compliance with policies and regulations. By adhering to these principles, organizations can create a robust data management framework that supports their strategic objectives.

10 Data Lifecycle Management

Data lifecycle management involves managing data from its creation to its eventual disposal, ensuring that it remains accurate, accessible, and secure throughout its lifecycle. The data lifecycle typically includes several stages: data creation and acquisition, data storage and maintenance, data usage and sharing, and data archiving and disposal. For example, during the data creation stage, data should be accurately captured and validated to prevent errors. In the storage and maintenance stage, data should be regularly updated and cleaned to maintain its accuracy and relevance. The usage and sharing stage involves ensuring that data is accessible to authorized users while protecting it from unauthorized access. Finally, the archiving and disposal stage involves securely storing data that is no longer actively used and disposing of data that is no longer needed, in compliance with retention policies.

11 Data Archiving and Retention

Data archiving and retention are critical components of data lifecycle management, ensuring that data is preserved and protected for as long as needed. Data archiving involves moving inactive data to a secure storage environment where it can be retained for long-term access. This helps free up space in active databases and improves system performance. Data retention policies define how long different types of data should be kept, based on legal, regulatory, and business requirements. Implementing these policies ensures that data is retained for the appropriate period and disposed of when it is no longer needed. Effective data archiving and retention also involve regularly reviewing and updating retention schedules to reflect changing regulations and business needs. By implementing robust data archiving and retention practices.

12 Regulatory Compliance and Risk Management

12.1 Ensuring Compliance with Data Laws

Ensuring compliance with data laws involves a comprehensive approach that includes establishing robust policies, implementing effective controls, and regularly monitoring and auditing data practices. Organizations should start by conducting a thorough assessment of their current data handling processes to identify areas of non-compliance. This involves mapping data flows, classifying data based on sensitivity, and understanding where and how data is stored and processed. Once potential gaps are identified, organizations must develop and implement policies that align with regulatory requirements, such as data access controls, data encryption, and regular data audits. Employee training is also crucial, ensuring that all staff members understand their responsibilities and the importance of data compliance. Regular audits and assessments help monitor compliance, allowing organizations to address any issues promptly and maintain adherence to evolving regulations.

12.2 Managing Data Privacy and Security

Data privacy and security are fundamental components of regulatory compliance and risk management. Data privacy involves protecting individuals' personal information from unauthorized access and ensuring that data is collected and used in a transparent and lawful manner. Key practices include obtaining explicit consent from individuals before collecting their data, providing clear privacy notices, and allowing individuals to access, correct, or delete their data. Data security focuses on protecting data from breaches, cyberattacks, and other threats. This involves implementing technical measures such as encryption, firewalls, and intrusion detection systems, as well as administrative measures like access controls, regular security assessments, and incident response plans. A robust data privacy and security framework not only ensures compliance with regulations but also builds customer trust and protects the organization from potential risks and liabilities.

12.3 Risk Assessment and Mitigation

Risk assessment and mitigation are critical processes in managing regulatory compliance and data security. A comprehensive risk assessment involves identifying potential threats to data, evaluating the likelihood and impact of these threats, and determining the organization's vulnerability to them. This can include threats such as data breaches, unauthorized access, data loss, and regulatory non-compliance. Once risks are identified, organizations must develop mitigation strategies to address them. These strategies can include implementing stronger security controls, such as multifactor authentication and advanced encryption, improving data governance practices, conducting regular security training for employees, and developing incident response plans to quickly address and mitigate any data breaches or compliance issues. Continuous monitoring and reassessment are also vital, as the threat landscape and regulatory requirements are constantly evolving. By proactively assessing and mitigating risks, organizations can protect their data, ensure compliance, and maintain a strong security posture.

In conclusion, regulatory compliance and risk management are essential components of effective data management. By understanding data regulations, ensuring compliance with data laws, managing data privacy and security, and conducting thorough risk assessments and mitigation, organizations can protect their data, avoid legal penalties, and maintain customer trust. Implementing a comprehensive data governance framework that incorporates these elements is critical for any organization that handles sensitive data in today's complex regulatory environment.

13 Leveraging Technology for Data Governance

13.1 Technology Solutions for Data Governance

Technology plays a pivotal role in modern data governance, offering solutions that enhance data management, security, and compliance. Technology solutions for data governance include data management platforms, governance frameworks, and data quality tools. These solutions help organizations establish consistent data practices, enforce policies, and automate governance processes. Key technology solutions include:

1. **Data Governance Platforms**: These platforms provide a centralized environment for managing data governance activities. They offer features such as metadata management, data lineage tracking, and policy enforcement. Examples include Collibra, Alation, and Informatica Axon.
2. **Data Catalogs**: Data catalogs enable organizations to inventory and organize their data assets. They provide a searchable repository of data, making it easier to understand data sources, usage, and ownership. Popular data catalog tools include Apache Atlas, Microsoft Azure Data Catalog, and IBM Watson Knowledge Catalog.
3. **Metadata Management Tools**: These tools help manage and maintain metadata, which provides context about data, such as its source, format, and usage. Effective metadata management supports data discovery, lineage, and impact analysis. Examples include erwin Data Intelligence and Talend Metadata Manager.
4. **Data Privacy Management Solutions**: Tools that help organizations manage data privacy compliance, including data mapping, consent management, and data subject access requests. Examples include OneTrust and TrustArc.

13.2 Data Quality Management Tools

Data quality management tools are essential for maintaining high data quality standards across an organization. These tools help identify, monitor, and resolve data quality issues such as inaccuracies, inconsistencies, and duplicates. Key data quality management tools include:

1. **Data Profiling Tools**: These tools analyze data to understand its structure, content, and quality. They help identify anomalies and data quality issues. Examples include Talend Data Profiler and IBM InfoSphere Information Analyzer.
2. **Data Cleansing Tools**: Tools designed to clean and standardize data by removing duplicates, correcting errors, and ensuring consistency. Examples include Trifacta and Data Ladder.
3. **Data Integration Tools**: These tools facilitate the integration of data from various sources, ensuring consistency and accuracy across systems. Examples include Informatica PowerCenter, MuleSoft, and Apache NiFi.

4. **Data Validation Tools**: Tools that apply validation rules to data to ensure it meets specified standards and formats. Examples include DataFlux and Oracle Enterprise Data Quality.
5. **Master Data Management (MDM) Solutions**: MDM solutions ensure that an organization maintains a single, consistent view of critical data across systems. Examples include IBM InfoSphere MDM, SAP Master Data Governance, and Informatica MDM.

13.3 Artificial Intelligence and Machine Learning in Data Governance

Artificial Intelligence (AI) and Machine Learning (ML) are transforming data governance by automating processes, improving data quality, and enhancing decision-making. AI and ML applications in data governance include:

1. **Automated Data Classification**: AI algorithms can automatically classify data based on its content and context, helping organizations apply appropriate governance policies and security measures. This reduces the need for manual data categorization and improves accuracy.
2. **Data Anomaly Detection**: Machine learning models can detect anomalies and outliers in data, identifying potential data quality issues or security breaches. These models learn from historical data and improve over time, providing more accurate detection capabilities.
3. **Predictive Data Quality Management**: AI and ML can predict potential data quality issues before they occur, allowing organizations to proactively address problems. This includes forecasting data degradation, identifying patterns, and recommending corrective actions.
4. **Natural Language Processing (NLP)**: NLP techniques enable better data understanding and management by extracting meaningful insights from unstructured data, such as text and documents. This helps in automating data discovery and enhancing metadata management.
5. **Enhanced Data Privacy**: AI-driven solutions can help manage data privacy by automating tasks such as data masking, encryption, and consent management. These solutions help ensure compliance with privacy regulations and protect sensitive data.

13.4 Big Data and Advanced Analytics

Big Data and advanced analytics play a crucial role in modern data governance by enabling organizations to handle and analyze vast amounts of data efficiently. Big Data technologies and analytics tools offer several benefits:

1. **Scalable Data Management**: Big Data platforms, such as Apache Hadoop and Apache Spark, provide scalable solutions for managing large volumes of data. These platforms enable organizations to process and analyze data from diverse sources quickly and efficiently.
2. **Real-Time Data Analytics**: Advanced analytics tools allow organizations to analyze data in real-time, providing actionable insights and enabling timely decision-making. Tools such as Apache Kafka and AWS Kinesis support real-time data processing and analysis.
3. **Data Visualization**: Big Data analytics tools offer advanced visualization capabilities, helping organizations to interpret complex data and communicate insights effectively. Tools such as Tableau, Power BI, and Qlik provide interactive dashboards and reports.
4. **Data Integration and Correlation**: Advanced analytics platforms facilitate the integration and correlation of data from multiple sources, providing a comprehensive view of data and enabling more accurate insights and decision-making.
5. **Predictive and Prescriptive Analytics**: These techniques use historical data and advanced algorithms to predict future trends and recommend actions. Predictive analytics helps forecast outcomes, while prescriptive analytics provides actionable recommendations based on data insights.

13.5 Integrating Technology into Data Governance Practices

Integrating technology into data governance practices involves aligning technological solutions with governance objectives and processes. Key steps include:

1. **Assessment of Needs**: Identify the specific data governance needs of the organization and select technology solutions that align with these needs. This involves evaluating existing systems, identifying gaps, and defining requirements.
2. **Technology Implementation**: Deploy selected technology solutions, ensuring they are properly configured and integrated with existing systems. This may involve setting up data governance platforms, integrating data quality tools, and implementing AI and ML models.
3. **Process Integration**: Integrate technology solutions into existing data governance processes, such as data quality management, data privacy, and risk assessment. Ensure that technology supports and enhances these processes rather than disrupting them.
4. **Training and Support**: Provide training and support to users to ensure they understand how to use technology solutions effectively. This includes training on new tools, processes, and best practices.
5. **Continuous Improvement**: Regularly review and update technology solutions to ensure they continue to meet evolving data governance needs. This involves monitoring performance, gathering user feedback, and making necessary adjustments.

6. **Collaboration and Communication**: Foster collaboration between technology teams and data governance stakeholders to ensure that technology solutions are effectively aligned with governance goals and requirements.

14 CaseStudy

XYZ Corporation (company Name is replaced with XYZ for privacy reason), a global leader in consumer electronics, faced significant challenges with its data management practices. The company struggled with data quality issues, regulatory compliance, and inefficient data governance processes. To address these challenges, XYZ Corporation decided to implement a comprehensive data governance technology solution that included a data governance platform, data quality management tools, and advanced analytics.

14.1 Objectives

1. Improve data quality and consistency across the organization.
2. Enhance compliance with data privacy regulations.
3. Streamline data governance processes and increase efficiency.
4. Leverage advanced analytics to gain actionable insights from data.

14.2 Implementation

XYZ Corporation implemented the following technology solutions:

1. **Data Governance Platform**: Collibra was chosen for its robust data governance capabilities, including metadata management, data lineage tracking, and policy enforcement.
2. **Data Quality Management Tools**: Informatica Data Quality and Trifacta were selected to address data profiling, cleansing, and validation needs.
3. **Advanced Analytics**: Tableau and Apache Spark were deployed for real-time data analysis and visualization.

The implementation process involved the following steps:

1. **Assessment and Planning**: Conducted a thorough assessment of existing data practices, identified gaps, and developed an implementation plan.
2. **Technology Deployment**: Configured and integrated Collibra, Informatica Data Quality, and Tableau into the organization's data management infrastructure.
3. **Training and Support**: Provided training to employees on new tools and processes, ensuring smooth adoption and effective usage.

4. **Monitoring and Evaluation**: Established metrics to measure the effectiveness of the technology solutions and monitored performance continuously.

14.3 *Quantitative Results*

The implementation of data governance technology at XYZ Corporation resulted in significant improvements across various metrics:

1. **Data Quality Improvement**:
   - **Error Rate Reduction**: The rate of data errors and inconsistencies decreased by 45% within the first six months of implementation. Prior to the implementation, the error rate was 12%, which dropped to 6.6% after deploying data quality management tools.
   - **Data Cleansing Efficiency**: Data cleansing processes were automated, reducing manual data correction efforts by 60%. Previously, data cleansing took an average of 120 h per month, which decreased to 48 h.
2. **Compliance and Privacy**:
   - **Regulatory Compliance**: The time required to prepare for regulatory audits decreased by 50%, from an average of 10 days to 5 days. This was achieved through improved data lineage tracking and automated compliance reporting.
   - **Data Privacy Incidents**: Data privacy incidents reduced by 70% due to enhanced data security measures and improved access controls. The number of reported incidents decreased from 20 per year to 6.
3. **Efficiency Gains**:
   - **Data Governance Process Efficiency**: The efficiency of data governance processes improved by 40%. The average time spent on data governance activities decreased from 40 h per week to 24 h per week, thanks to automated workflows and better data visibility.
   - **Decision-Making Speed**: The time required to generate data-driven reports and insights decreased by 35%, from 48 to 31 h, due to the deployment of advanced analytics tools and real-time data processing.
4. **Advanced Analytics**:
   - **Insight Generation**: The ability to generate actionable insights from data improved by 50%. The number of actionable insights derived from data increased from 15 per month to 22 per month, enabling better strategic decision-making.
   - **Data Visualization Adoption**: User adoption of data visualization tools increased by 80%. The number of users actively using Tableau for data analysis and reporting grew from 50 to 90.

The implementation of data governance technology at XYZ Corporation led to substantial improvements in data quality, regulatory compliance, and operational efficiency. By leveraging a comprehensive data governance platform, data quality management tools, and advanced analytics, XYZ Corporation was able to address its data management challenges effectively. The quantitative results demonstrate the positive impact of technology on data governance practices, highlighting the benefits of investing in advanced data management solutions. The case study underscores the importance of technology in enhancing data governance and achieving strategic business objectives.

Table 1 provides a clear overview of the quantitative results observed at XYZ Corporation after implementing data governance technology solutions and comparative analysis before and after implementation is shown in Fig. 3. The improvements span various aspects of data management, including data quality, compliance, efficiency, and advanced analytics.

Table 1 Case study results: implementing data governance technology

Metric	Before implementation	After implementation	Percentage improvement (%)
Data quality improvement			
Error rate	12%	6.6%	45
Data cleansing effort (Hours/Month)	120 h	48 h	60
Compliance and privacy			
Time for regulatory audits (Days)	10 days	5 days	50
Data privacy incidents (Per year)	20	6	70
Efficiency gains			
Data governance process efficiency (%)	Baseline not available	Improved by 40%	–
Time to generate reports (Hours)	48 h	31 h	35
Advanced analytics			
Number of actionable insights (Per month)	15	22	50
Data visualization tool adoption (Users)	50	90	80

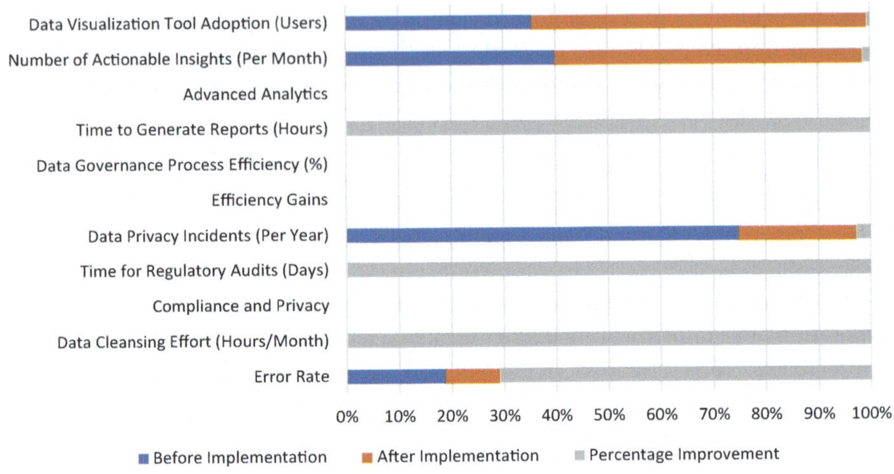

Fig. 3 Comparative analysis before and after implementation

15 Conclusion and Future Work

The implementation of advanced data governance technologies at XYZ Corporation has yielded significant improvements in data quality, regulatory compliance, and operational efficiency. By integrating a robust data governance platform, deploying data quality management tools, and utilizing advanced analytics, XYZ Corporation has successfully addressed key challenges related to data management. The marked reduction in data errors, enhanced compliance with data privacy regulations, and increased efficiency in data governance processes underscore the effectiveness of these technological solutions. Additionally, the advancements in data analytics have enabled the organization to generate more actionable insights and make data-driven decisions more swiftly, thereby supporting strategic business objectives.

Looking ahead, there are several areas for future work to further enhance data governance practices. As data environments continue to evolve, incorporating emerging technologies such as blockchain for data integrity and more advanced AI algorithms for predictive analytics could offer new opportunities for improving data governance. Furthermore, expanding the scope of data governance to include new data sources and types, such as unstructured data from social media and IoT devices, will be crucial for maintaining comprehensive data oversight. Continuous evaluation and adaptation of technology solutions will be necessary to address the dynamic nature of data privacy regulations and emerging security threats. By staying ahead of technological advancements and regulatory changes, organizations can ensure that their data governance frameworks remain robust and capable of supporting their evolving business needs.

References

1. Singh, K., Li, S., Jahnke, I., Alarcon, M.L., Mosa, A., Calyam, P.: Improving big data governance in healthcare institutions: user experience research for honest broker based application to access healthcare big data. Behav. Inform. Technol. **43**(6), 1067–1095 (2024)
2. Gao, J., Sun, Y., Rameezdeen, R., Chow, C.: Understanding data governance requirements in IoT adoption for smart ports–a gap analysis. Marit. Policy Manag. **51**(4), 617–630 (2024)
3. Pearlson, K.E., Saunders, C.S., Galletta, D.F.: Managing and Using Information Systems: A Strategic Approach. John Wiley & Sons (2024)
4. Al Wahid, S.A., Mohammad, N., Islam, R., Faisal, M.H., Rana, M.S.: Evaluation of information technology implementation for business goal improvement under process functionality in economic development. J. Data Anal. Inform. Process. **12**(2), 304–317 (2024)
5. Sonmez Cakir, F., Kalaycioglu, O., Adiguzel, Z.: Examination the effects of organizational innovation and knowledge management strategy in information technology companies in R&D departments on service quality and product innovation. Inf. Technol. People **37**(4), 1540–1559 (2024)
6. Ochuba, N.A., Amoo, O.O., Okafor, E.S., Akinrinola, O., Usman, F.O.: Strategies for leveraging big data and analytics for business development: a comprehensive review across sectors. Comp. Sci. IT Res. J. **5**(3), 562–575 (2024)
7. Oriji, O., Joel, O.S.: Integrating accounting models with supply chain management in the aerospace industry: a strategic approach to enhancing efficiency and reducing costs in the US. World J. Adv. Res. Rev. **21**(3), 1476–1489 (2024)
8. Bernardi, F.A.: Digital health research governance: from FAIR to RE-AIM. Doctoral dissertation, Universidade de São Paulo (2024)
9. Bammidi, T.R., Gutta, L.M., Kotagiri, A., Samayamantri, L.S., Krishna Vaddy, R.: The crucial role of data quality in automated decision-making systems. Int. J. Manage. Educ. Sustain. Dev. **7**(7), 1–22 (2024)
10. Whig, P., Silva, N., Elngar, A.A., Aneja, N., Sharma, P. (eds.): Sustainable Development through Machine Learning, AI and IoT: First International Conference, ICSD 2023, Delhi, India, July 15–16, 2023, Revised Selected Papers. Springer Nature (2023)
11. Yandrapalli, V.: AI-powered data governance: a cutting-edge method for ensuring data quality for machine learning applications. In: 2024 Second International Conference on Emerging Trends in Information Technology and Engineering (ICETITE), pp. 1–6. IEEE (2024, February)
12. Channa, A., Sharma, A., Singh, M., Malhotra, P., Bajpai, A., Whig, P.: Original research article revolutionizing filmmaking: a comparative analysis of conventional and AI-generated film production in the era of virtual reality. J. Autonom. Intell. **7**(4) (2024)
13. Moinuddin, M., Usman, M., Khan, R.: Strategic insights in a data-driven era: maximizing business potential with analytics and AI. Revista Espanola de Documentacion Cientifica **18**(02), 117–133 (2024)
14. Shafiq, W.: Optimizing organizational performance: a data-driven approach in management science. Bull. Manage. Rev. **1**(2), 31–40 (2024)
15. Jain, A., Kamat, S., Saini, V., Singh, A., Whig, P.: Agile leadership: navigating challenges and maximizing success. In: Practical Approaches to Agile Project Management, pp. 32–47. IGI Global (2024)
16. Whig, P., Remala, R., Mudunuru, K.R., Quraishi, S.J.: Integrating AI and quantum technologies for sustainable supply chain management. In: Quantum Computing and Supply Chain Management: A New Era of Optimization, pp. 267–283. IGI Global (2024)
17. Mittal, S., Koushik, P., Batra, I., Whig, P.: AI-driven inventory management for optimizing operations with quantum computing. In: Quantum Computing and Supply Chain Management: A New Era of Optimization, pp. 125–140. IGI Global (2024)
18. Whig, P., Mudunuru, K.R., Remala, R.: Quantum-inspired data-driven decision making for supply chain logistics. In: Quantum Computing and Supply Chain Management: A New Era of Optimization, pp. 85–98. IGI Global (2024)

19. Sehrawat, S.K., Dutta, P.K., Bhatia, A.B., Whig, P.: Predicting demand in supply chain networks with quantum machine learning approach. In: Quantum Computing and Supply Chain Management: A New Era of Optimization, pp. 33–47. IGI Global (2024)
20. Whig, P., Kasula, B.Y., Yathiraju, N., Jain, A., Sharma, S.: Transforming aviation: the role of artificial intelligence in air traffic management. In: New Innovations in AI, Aviation, and Air Traffic Technology, pp. 60–75. IGI Global (2024)
21. Kasula, B.Y., Whig, P., Vegesna, V.V., Yathiraju, N.: Unleashing exponential intelligence: transforming businesses through advanced technologies. Int. J. Sustain. Dev. Through AI, ML IoT **3**(1), 1–18 (2024)
22. Whig, P., Bhatia, A.B., Nadikatu, R.R., Alkali, Y., Sharma, P.: 3 security issues. In: Software-Defined Network Frameworks: Security Issues and Use Cases, 34 (2024)
23. Pansara, R.R., Mourya, A.K., Alam, S.I., Alam, N., Yathiraju, N., Whig, P.: Synergistic integration of master data management and expert system for maximizing knowledge efficiency and decision-making capabilities. In: 2024 2nd International Conference on Advancement in Computation & Computer Technologies (InCACCT), pp. 13–16. IEEE (2024, May)
24. Whig, P., Kautish, S.: VUCA leadership strategies models for pre-and post-pandemic scenario. In: VUCA and Other Analytics in Business Resilience, Part B, pp. 127–152. Emerald Publishing Limited (2024)
25. Whig, P., Bhatia, A.B., Nadikatu, R.R., Alkali, Y., Sharma, P.: GIS and remote sensing application for vegetation mapping. In: Geo-Environmental Hazards using AI-enabled Geospatial Techniques and Earth Observation Systems, pp. 17–39. Springer Nature Switzerland, Cham (2024)

Strategic Analytics for Startup Organizations

Pawan Whig, Nikitha Yathiraju, Anupriya Jain, Seema Sharma, and Sandeep Kautish

Abstract In the dynamic landscape of startup ecosystems, strategic decision-making is paramount for sustainable growth and competitive advantage. This chapter explores the integration of analytics into the strategic framework of startup organizations. It delves into the methodologies, tools, and frameworks that startups can leverage to harness data-driven insights for informed decision-making. Key topics include the identification of strategic goals, data collection and management strategies, analytical techniques for market intelligence and customer segmentation, and the implementation of analytics to optimize operations and enhance scalability. By adopting a strategic analytics approach, startups can navigate uncertainties more effectively, capitalize on emerging opportunities, and chart a path towards long-term success in a rapidly evolving market environment.

Keywords Strategic analytics · Startup organizations · Data-driven insights · Decision-making · Competitive advantage · Market intelligence · Customer segmentation

P. Whig (✉)
VIPS-TC, New Delhi, India
e-mail: pawanwhig@gmail.com

N. Yathiraju
University of the Cumberland, Williamsburg, USA

A. Jain · S. Sharma
SCA, MRIIRS, Faridabad, India

S. Kautish
Chandigarh University, Mohali, India

S. Kautish et al. (eds.), *Strategy Analytics for Business Resilience Theories and Practices*, Information Systems Engineering and Management 33,
https://doi.org/10.1007/978-3-031-82369-5_2

1 Introduction

In today's competitive business landscape, startups face a myriad of challenges and opportunities. From navigating market uncertainties to scaling operations efficiently, the path to success is often paved with strategic decisions that can make or break a young company. Strategic analytics emerges as a powerful toolset that startups can leverage to not only survive but thrive in this dynamic environment.

At its core, strategic analytics in startups revolves around harnessing data-driven insights to inform and validate strategic decisions. Unlike established corporations with vast resources, startups often operate under tighter constraints, making every decision critical to their survival and growth. By leveraging analytics, these companies can gain a deeper understanding of their market, customers, and operational efficiencies, thereby optimizing their limited resources for maximum impact. One of the foundational elements of strategic analytics is aligning data analysis efforts with the startup's strategic goals. These goals typically include achieving product-market fit, scaling operations, and establishing a competitive edge in the market. For instance, a startup aiming to disrupt a traditional industry with a new technology must use analytics to identify market gaps, understand customer pain points, and validate the demand for their innovation.

Effective strategic analytics begins with robust data collection and management strategies. Startups often have access to a wealth of data, ranging from user interactions on their platforms to market trends and competitor activities. However, the challenge lies in aggregating, cleansing, and storing this data in a manner that facilitates meaningful analysis. Cloud-based solutions and data management platforms offer scalable options for startups to centralize their data repositories securely. Once data is collected and organized, startups can employ various analytical techniques to extract actionable insights. Market intelligence, for example, involves analyzing external market data to identify trends, opportunities, and threats. Startups can use techniques such as regression analysis, time series forecasting, and sentiment analysis to anticipate market shifts and adapt their strategies accordingly.

Understanding customer behavior is another critical area where strategic analytics can drive significant value for startups. By segmenting customers based on demographics, behaviors, or preferences, startups can tailor their marketing efforts and product offerings to better meet customer needs. Advanced analytics, including clustering algorithms and predictive modeling, enable startups to create personalized experiences that enhance customer satisfaction and loyalty. Operational efficiency is paramount for startups aiming to scale rapidly without compromising quality or sustainability. Strategic analytics plays a pivotal role in optimizing internal processes, identifying bottlenecks, and streamlining workflows. Techniques such as process mining and performance analytics allow startups to pinpoint inefficiencies and implement data-driven improvements that drive operational excellence. Literature Review in tabular form with research gap is shown in Table 1.

Table 1 Literature review with research gap

Citation	Title	Key findings	Research gap
Joel et al. [1]	Navigating the digital transformation journey: strategies for startup growth and innovation in the digital era	Digital transformation strategies for startup growth	Lack of specific case studies on different types of startups
Ochuba et al. [2]	Innovations in business models through strategic analytics and management	Use of strategic analytics for sustainable growth in businesses	Need for empirical data on long-term outcomes of strategic analytics in startups
Zamani et al. [3]	Making sense of business analytics in project selection and prioritisation	Insights on project selection using business analytics	Application of these insights across different industries and geographies
Ibeh et al. [4]	Business analytics and decision science: A review of techniques in strategic business decision making	Review of business analytics techniques for strategic decision-making	Integration of advanced AI techniques with traditional business analytics
Timóteo et al. [5]	Use of AHP and grey fixed weight clustering to assess maturity level of strategic communication management in Brazilian startups	Assessment of strategic communication maturity in startups using AHP and grey clustering	Exploration of other strategic communication assessment tools
Martins [6]	Startup Guide to AI: Integrating Technology for Business Success	Guide on AI integration for business success in startups	Need for longitudinal studies to assess the impact of AI integration over time
Chaudhri et al. [7]	Assembling the Start-up Brand: a process framework for understanding strategic communication challenges	Framework for understanding strategic communication challenges in startups	Examination of the framework's applicability in diverse cultural settings
Kim [8]	Startup acquisitions as a hiring strategy: Turnover differences between acquired and regular hires	Differences in turnover rates between acquired and regular hires in startups	Comparative studies on turnover impacts across various acquisition strategies
Ajalkar et al. [9]	Futuristic business intelligence framework for start-ups	Proposal of a futuristic BI framework for startups	Validation of the framework through real-world implementation

(continued)

Table 1 (continued)

Citation	Title	Key findings	Research gap
Tan et al. [10]	Unlocking venture growth: Synergizing big data analytics, artificial intelligence, new product development practices, and inter-organizational digital capability	Synergy of big data analytics, AI, and digital capabilities for venture growth	In-depth studies on the interaction between these elements in different sectors
Bandini [11]	The business development of a Startup and the market study of an AI innovative technology	Market study on AI technology in startups	Broader market studies across multiple regions and sectors
Rudeloff et al. [12]	Conditions of one-way and two-way approaches in strategic start-up communication: a qualitative comparative analysis (QCA)	Analysis of one-way and two-way communication in startups	Quantitative studies to support the qualitative findings
York et al. [13]	Lean startup and learning loops in entrepreneurial ventures: a systematic review	Systematic review of lean startup and learning loops	Practical application and effectiveness of Lean Startup principles in various industries
Agarwal and Sharma [14]	Digital marketing of startup businesses	Digital marketing strategies for startups	Exploration of evolving digital marketing trends and their effectiveness
Grover and Piccoli [15]	A foundational perspective of digital business strategy for strategic leaders	Perspective on digital business strategy for leaders	Detailed case studies on the implementation of these strategies in startups
Kamboj [16]	Role of technology in fostering innovation and growth in startup businesses	Role of technology in driving innovation and growth in startups	Longitudinal studies on the sustained impact of technology on innovation
Olajiga et al. [17]	Data analytics in energy corporations: Conceptual framework for strategic business outcomes	Framework for data analytics in energy corporations	Adaptation of the framework to startups in other industries

2 Navigating Uncertainties and Seizing Opportunities

In the volatile startup ecosystem, uncertainties abound, from regulatory changes to economic downturns and technological disruptions. Strategic analytics provides startups with the agility to navigate these uncertainties by continuously monitoring key performance indicators (KPIs), conducting scenario analyses, and developing contingency plans. Moreover, analytics helps startups identify emerging opportunities in the market, enabling them to capitalize on first-mover advantages and innovate proactively. Ultimately, the goal of strategic analytics for startups extends beyond short-term gains to sustainable long-term success. By fostering a culture of data-driven decision-making, startups can foster innovation, iterate on their products/services based on real-time feedback, and build a resilient business model. Continuous improvement through analytics allows startups to stay ahead of competitors, attract investment, and forge lasting relationships with customers and stakeholders. As startups continue to disrupt industries and drive economic growth, the importance of strategic analytics cannot be overstated. By embracing data-driven insights, startups can navigate complexities with confidence, mitigate risks, and seize opportunities that propel them towards sustainable growth and market leadership. This chapter will delve deeper into these themes, exploring case studies, best practices, and emerging trends that showcase the transformative power of strategic analytics in startup organizations.

2.1 Data Collection

Data collection is the foundational step in building a robust analytics framework for startups. It involves gathering relevant information from various sources to support informed decision-making and insights generation. For startups, the challenge lies not only in collecting data but also in ensuring its relevance, accuracy, and compliance with privacy regulations. Here are some key data collection methods startups can utilize:

2.1.1 Direct Interaction and Feedback

Customer Surveys and Feedback Forms: Startups can gather direct insights from customers through surveys, feedback forms on websites or apps, and social media polls. These methods provide qualitative data on customer preferences, satisfaction levels, and pain points.

Interviews and Focus Groups: Conducting interviews and focus groups allows startups to delve deeper into specific topics, gaining qualitative insights that may not be captured through other methods.

2.1.2 Transactional Data

Sales and Purchase Data: Startups can analyze transactional data from sales records, invoices, and payment processing systems. This data provides insights into product performance, customer buying patterns, and revenue generation.

Web Analytics: Utilizing tools like Google Analytics or Adobe Analytics helps startups track website traffic, user behavior, and conversion rates. This data is crucial for optimizing digital marketing strategies and improving user experience.

2.1.3 Operational Data

Internal Systems and Processes: Startups can leverage data from operational systems such as CRM (Customer Relationship Management), ERP (Enterprise Resource Planning), and project management tools. This data helps optimize internal processes, resource allocation, and project timelines.

2.1.4 External Data Sources

Market Research Reports: Accessing industry reports, competitor analysis, and market research studies provides startups with external market data. This information aids in understanding market trends, identifying opportunities, and benchmarking against competitors.

Publicly Available Data: Startups can use publicly available data from government databases, open data initiatives, and APIs (Application Programming Interfaces) to enrich their datasets. This data may include demographic information, economic indicators, or environmental data relevant to business operations.

2.1.5 Sensor and IoT Data

IoT Devices and Sensors: For startups involved in hardware or IoT (Internet of Things) solutions, data from sensors and connected devices offer real-time insights into product performance, usage patterns, and environmental conditions.

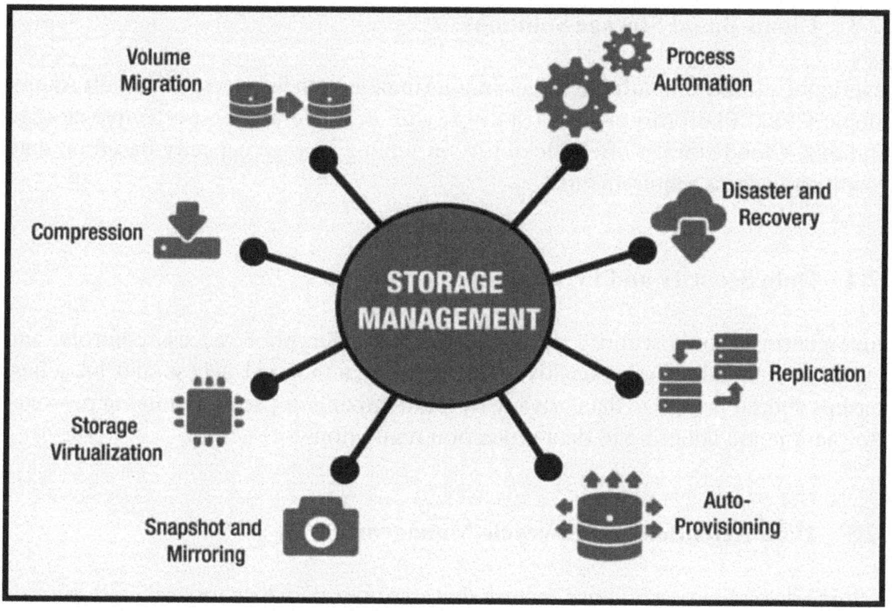

Fig. 1 Storage management solutions

2.2 *Data Management and Storage Solutions*

Effective data management and storage as shown in Fig. 1 are essential for startups to derive actionable insights from their collected data while ensuring security, accessibility, and scalability. Startups must consider the following aspects when developing their data management strategies.

2.2.1 Data Governance

Establishing clear policies and procedures for data collection, usage, and access ensures consistency and compliance with regulatory requirements (e.g., GDPR, CCPA). Startups should designate roles and responsibilities for data stewardship and implement mechanisms for data quality assurance.

2.2.2 Data Integration and Aggregation

Integrating data from disparate sources into a centralized repository facilitates comprehensive analysis and reporting. Startups can use ETL (Extract, Transform, Load) tools or data integration platforms to automate the process of aggregating and harmonizing data.

2.2.3 Cloud-Based Storage Solutions

Leveraging cloud computing services (e.g., Amazon Web Services, Microsoft Azure, Google Cloud Platform) provides startups with scalable and cost-effective storage solutions. Cloud storage offers flexibility in scaling storage capacity based on data growth and access requirements.

2.2.4 Data Security and Privacy

Implementing robust security measures, such as encryption, access controls, and regular audits, safeguards sensitive data from unauthorized access and breaches. Startups should prioritize data privacy by anonymizing or pseudonymizing personal information and adhering to data protection regulations.

2.2.5 Data Retention and Lifecycle Management

Defining data retention policies ensures that startups retain data for necessary periods while adhering to legal and business requirements. Implementing data lifecycle management practices involves archiving obsolete data and managing data expiration to optimize storage resources.

2.2.6 Analytics and Visualization Tools

Deploying analytics and visualization tools (e.g., Tableau, Power BI, Looker) enables startups to explore data insights through interactive dashboards and reports. These tools empower stakeholders to derive actionable insights quickly and make data-driven decisions.

2.2.7 Scalability and Performance

Designing data management solutions that can scale with business growth and handle increasing volumes of data ensures continuous performance and reliability. Startups should monitor system performance metrics and optimize infrastructure as needed to maintain responsiveness.

By adopting robust data collection methods and implementing effective data management and storage solutions, startups can harness the power of data analytics to drive innovation, optimize operations, and achieve sustainable growth in competitive markets. These foundational practices lay the groundwork for building a data-driven culture that prioritizes informed decision-making and agility in responding to market dynamics.

3 Regression Analysis and Forecasting

Regression Analysis: Regression analysis is a statistical technique used to examine the relationship between one dependent variable (often the outcome of interest) and one or more independent variables (predictors). In the context of market intelligence for startups, regression analysis helps to:

- **Identify Relationships**: Startups can use regression to identify and quantify relationships between variables such as marketing spend and sales revenue, pricing and customer demand, or product features and user engagement.
- **Predict Outcomes**: By analyzing historical data, startups can build regression models to predict future outcomes. For example, a startup can predict how changes in advertising expenditure will impact sales based on past data patterns.
- **Optimize Decision-making**: Regression analysis provides insights into which factors influence key business metrics, enabling startups to optimize strategies for pricing, marketing campaigns, and resource allocation.

Forecasting: Forecasting involves using historical data to predict future trends and outcomes. Startups can employ various forecasting techniques such as time series analysis, moving averages, and exponential smoothing to:

- **Predict Demand**: Startups can forecast future demand for their products or services based on historical sales data, seasonal variations, and market trends. This helps in inventory management, production planning, and resource allocation.
- **Financial Planning**: Forecasting financial metrics such as revenue, expenses, and cash flow assists startups in budgeting, fundraising, and financial decision-making.
- **Scenario Analysis**: By creating multiple forecasts under different scenarios (e.g., optimistic, pessimistic), startups can assess the potential impact of various market conditions and develop contingency plans.

4 Sentiment Analysis and Customer Insights

Sentiment Analysis: Sentiment analysis, also known as opinion mining, involves analyzing text data (such as customer reviews, social media posts, or survey responses) to determine the sentiment expressed towards a product, brand, or service. For startups, sentiment analysis provides valuable insights:

- **Customer Feedback Analysis**: Startups can analyze customer reviews on platforms like Amazon, Yelp, or social media to understand overall sentiment and identify specific areas for improvement.
- **Brand Reputation Management**: Monitoring sentiment helps startups track brand perception over time, identify emerging issues or trends, and take proactive measures to manage reputation.

- **Competitor Analysis**: Sentiment analysis can be used to compare customer sentiment towards competitors' products or services, providing insights into competitive strengths and weaknesses.

Customer Insights: Customer insights refer to actionable information derived from analyzing customer data, behaviors, and preferences. Startups can leverage customer insights to:

- **Segmentation and Targeting**: By segmenting customers based on demographic, behavioral, or psychographic factors, startups can tailor marketing campaigns and product offerings to specific customer segments.
- **Personalization**: Understanding customer preferences enables startups to personalize the user experience, recommend relevant products/services, and improve customer satisfaction and loyalty.
- **Predictive Analytics**: By combining customer insights with predictive modeling techniques, startups can anticipate future behavior, such as likelihood to purchase, churn risk, or product adoption rates.

Leveraging regression analysis and forecasting enhances startups' ability to make data-driven predictions and optimize strategic decisions. Similarly, employing sentiment analysis and customer insights enables startups to understand customer sentiment, improve brand perception, and drive personalized marketing strategies. By integrating these analytical techniques into their market intelligence efforts, startups can gain a competitive edge, innovate effectively, and foster sustainable growth in dynamic market environments.

5 Demographic and Behavioral Segmentation

Demographic Segmentation: Demographic segmentation divides customers into groups based on identifiable characteristics such as age, gender, income, education, occupation, and family status. For startups, demographic segmentation provides insights into:

- **Target Audience Identification**: By analyzing demographic data, startups can identify their primary target audience and tailor marketing messages and product features to appeal to specific demographic groups.
- **Market Sizing and Opportunity Assessment**: Understanding the size and characteristics of different demographic segments helps startups assess market potential, prioritize customer acquisition efforts, and allocate resources effectively.
- **Localization and Cultural Relevance**: Demographic segmentation enables startups to localize their offerings by considering cultural preferences, language nuances, and regional differences among target demographics.

Behavioral Segmentation: Behavioral segmentation categorizes customers based on their actions, behaviors, and usage patterns. Key behavioral segmentation variables include purchase history, frequency of purchases, brand loyalty, engagement with marketing campaigns, and usage of products/services. For startups, behavioral segmentation offers several benefits:

- **Predictive Modeling**: Analyzing past behaviors helps startups predict future actions, such as likelihood to purchase, churn risk, or product adoption rates. This enables personalized marketing strategies and proactive customer retention efforts.
- **Customer Journey Mapping**: Understanding customer behaviors throughout the buying journey helps startups optimize touchpoints, improve user experience, and address pain points that impact conversion rates.
- **Segment-Specific Campaigns**: Startups can design targeted marketing campaigns and promotions tailored to each behavioral segment's preferences, interests, and buying habits. This increases campaign effectiveness and ROI.

6 Personalization Strategies Using Analytics

Data-driven Customer Profiling: Personalization begins with creating detailed customer profiles based on demographic, behavioral, and psychographic data. Startups can leverage analytics to:

- **Aggregate and Analyze Data**: Integrate customer data from multiple sources (e.g., CRM systems, transaction records, website interactions) to build comprehensive customer profiles. Analyze this data to uncover patterns and preferences that inform personalized recommendations.
- **Segment-based Personalization**: Tailor product recommendations, marketing messages, and content based on customer segment characteristics. Use analytics to dynamically adjust personalization strategies based on real-time customer interactions and feedback.

Dynamic Content Personalization: Dynamic content personalization involves delivering customized content, offers, and experiences in real-time based on individual customer behavior and preferences. Startups can implement:

- **Recommendation Engines**: Utilize collaborative filtering algorithms and machine learning models to suggest products or content relevant to each customer's interests and past interactions.
- **Behavioral Triggers**: Set up automated triggers based on customer actions (e.g., abandoned cart reminders, personalized email campaigns triggered by specific behaviors) to drive engagement and conversions.

Predictive Personalization: Predictive analytics enables startups to anticipate customer needs and preferences before they are explicitly expressed. By applying predictive models, startups can:

- **Anticipate Customer Intent**: Predict customer behavior and preferences based on historical data, current trends, and contextual factors. Use these insights to proactively personalize interactions and offerings.
- **Lifecycle Stage Personalization**: Tailor communications and offerings based on where customers are in their lifecycle (e.g., new customer onboarding, loyal customer rewards, win-back campaigns for inactive customers).

Customer segmentation allows startups to categorize their audience into distinct groups based on demographics and behaviors, enabling targeted marketing and personalized experiences. By leveraging analytics for personalization, startups can create meaningful customer relationships, drive engagement, and enhance overall satisfaction, ultimately contributing to long-term customer loyalty and business growth.

7 Process Mining and Workflow Optimization

Process Mining: Process mining is a data-driven methodology that involves analyzing event logs and data from operational systems to discover, monitor, and improve real processes. For startups, process mining provides insights into:

- **Process Discovery**: By analyzing event logs from systems like ERP, CRM, or workflow management tools, startups can visualize and map out their business processes. This helps identify bottlenecks, inefficiencies, and deviations from intended workflows.
- **Root Cause Analysis**: Process mining enables startups to pinpoint the root causes of operational issues or delays. By identifying where and why processes break down, startups can implement targeted improvements to enhance efficiency and throughput.
- **Compliance Monitoring**: Startups can ensure compliance with internal policies, regulatory requirements, and industry standards by monitoring process adherence and identifying deviations that require corrective action.

Workflow Optimization: Based on insights gained from process mining, startups can optimize workflows to streamline operations and improve overall performance:

- **Process Redesign**: Redesigning workflows based on process mining findings allows startups to simplify complex processes, eliminate redundant steps, and optimize resource allocation.
- **Automation Opportunities**: Identifying repetitive tasks or manual interventions that can be automated enhances operational efficiency and reduces error rates. Startups can leverage workflow automation tools and RPA (Robotic Process Automation) solutions to streamline operations.

- **Continuous Improvement**: Implementing a culture of continuous improvement ensures that workflow optimizations are iterative and responsive to evolving business needs. Startups can use feedback loops and performance metrics to monitor the impact of workflow changes and make further refinements.

8 Performance Analytics and KPI Monitoring

Performance Analytics: Performance analytics involves measuring and analyzing key metrics to evaluate the effectiveness and efficiency of business operations. For startups, performance analytics provides insights into:

- **Operational Efficiency**: Tracking metrics such as cycle time, throughput, and resource utilization helps startups identify opportunities for efficiency gains and operational cost savings.
- **Quality and Compliance**: Monitoring quality metrics and compliance indicators ensures that startups deliver products or services that meet customer expectations and regulatory requirements.
- **Customer Satisfaction**: Analyzing customer service metrics (e.g., response time, resolution rates) and customer feedback helps startups gauge satisfaction levels and identify areas for service improvement.

KPI Monitoring: Key Performance Indicators (KPIs) are quantifiable metrics that measure the success of specific business objectives. Startups can monitor KPIs to:

- **Strategic Alignment**: Align operational activities with strategic goals by tracking KPIs related to revenue growth, market share, customer acquisition costs, and profitability.
- **Real-time Decision-making**: Utilize real-time KPI dashboards and reports to monitor performance trends, identify anomalies, and make data-driven decisions promptly.
- **Benchmarking and Goal Setting**: Compare current performance against industry benchmarks and historical data to set achievable goals and milestones for continuous improvement.

Operational optimization and scalability are critical for startups aiming to maximize efficiency, minimize costs, and maintain competitiveness. By leveraging process mining for workflow optimization and implementing performance analytics with KPI monitoring, startups can streamline operations, enhance decision-making capabilities, and foster a culture of continuous improvement that supports sustainable growth in dynamic market environments.

9 Scenario Analysis and Risk Management

Scenario Analysis: Scenario analysis involves creating and analyzing multiple hypothetical scenarios to assess the potential impact of various future events or conditions on business outcomes. For startups, scenario analysis serves several purposes:

- **Risk Assessment**: Startups can identify and prioritize potential risks by exploring different scenarios, such as economic downturns, regulatory changes, or technological disruptions.
- **Contingency Planning**: By preparing contingency plans for each scenario, startups can mitigate risks, minimize potential losses, and maintain operational continuity in adverse conditions.
- **Strategic Decision-making**: Scenario analysis helps startups anticipate challenges and opportunities, enabling proactive decision-making and adaptive strategies based on anticipated future conditions.

Risk Management: Risk management involves identifying, assessing, and mitigating risks that could impact business objectives. Startups can implement effective risk management practices such as:

- **Risk Identification**: Conducting thorough risk assessments to identify internal and external risks that could affect business operations, finances, reputation, or compliance.
- **Risk Mitigation Strategies**: Developing risk mitigation strategies tailored to specific risks, such as diversifying suppliers, implementing cybersecurity measures, or securing insurance coverage.
- **Risk Monitoring and Reporting**: Continuously monitoring key risk indicators (KRIs) and implementing early warning systems to detect emerging risks promptly. Regular risk reporting facilitates informed decision-making and transparency across the organization.

10 Identifying and Capitalizing on Emerging Opportunities

Market Monitoring and Trend Analysis: Startups can identify emerging opportunities by monitoring market trends, consumer behavior shifts, and technological advancements:

- **Market Research**: Conducting ongoing market research and competitor analysis helps startups stay informed about industry trends, customer preferences, and competitive landscape changes.
- **Early Adopter Advantage**: Identifying emerging technologies or market niches allows startups to position themselves as early adopters, gaining a competitive edge and capturing new market segments.

- **Agile Innovation**: Adopting agile innovation practices enables startups to rapidly prototype, test, and iterate new products or services in response to emerging opportunities and market demands.

Strategic Partnerships and Networking: Building strategic partnerships and networks with industry experts, investors, and potential collaborators enhances startups' access to market insights and partnership opportunities:

- **Collaborative Innovation**: Partnering with industry leaders or research institutions facilitates collaborative innovation, knowledge sharing, and access to specialized resources or technologies.
- **Investment and Funding Opportunities**: Engaging with investors and venture capitalists who specialize in emerging technologies or industries can provide startups with funding and strategic guidance to capitalize on growth opportunities.
- **Adaptability and Flexibility**: Maintaining organizational adaptability and flexibility allows startups to pivot quickly in response to changing market dynamics and capitalize on emerging opportunities as they arise.

Navigating uncertainties and seizing opportunities are integral to startup success in dynamic and competitive markets. By conducting scenario analysis for risk management, startups can anticipate and mitigate potential risks while identifying and capitalizing on emerging opportunities through proactive market monitoring, strategic partnerships, and agile innovation practices. Embracing a proactive and adaptive approach enables startups to navigate uncertainties effectively and position themselves for sustainable growth and market leadership.

11 Cultivating a Data-Driven Culture

Definition and Importance: Cultivating a data-driven culture involves fostering an organizational mindset where data is used systematically to inform decision-making across all levels. For startups, establishing a data-driven culture is crucial for:

- **Informed Decision-making**: Encouraging stakeholders to base decisions on data-driven insights rather than intuition or anecdotal evidence. This ensures decisions are grounded in empirical evidence and aligned with strategic objectives.
- **Transparency and Accountability**: Promoting transparency by making data accessible to all relevant stakeholders. This fosters accountability as decisions can be traced back to data-backed rationale rather than subjective opinions.
- **Learning and Adaptation**: Creating a learning environment where continuous data analysis and feedback loops facilitate adaptive strategies and rapid response to market changes.

Key Components: To cultivate a data-driven culture, startups can focus on several key components:

- **Leadership Support**: Leadership endorsement and commitment to data-driven practices set the tone for the entire organization. Leaders should prioritize data literacy, allocate resources for data initiatives, and lead by example in using data for decision-making.
- **Data Literacy and Training**: Providing training and development opportunities to enhance data literacy among employees. This includes understanding data analytics tools, interpreting data visualizations, and applying analytical techniques to business challenges.
- **Data Accessibility**: Ensuring data accessibility through centralized data repositories, user-friendly analytics platforms, and clear data governance policies. Empowering employees with self-service analytics capabilities encourages exploration and experimentation with data.
- **Collaboration and Communication**: Encouraging cross-functional collaboration where data scientists, analysts, and business stakeholders work together to derive insights and co-create solutions. Effective communication of data findings and implications ensures alignment and informed decision-making.

12 Continuous Improvement and Innovation

Continuous Improvement: Continuous improvement involves ongoing efforts to enhance products, processes, and organizational capabilities based on data-driven insights and stakeholder feedback. For startups, continuous improvement fosters:

- **Operational Efficiency**: Identifying inefficiencies, optimizing workflows, and reducing waste through iterative improvements informed by performance analytics and process optimization.
- **Quality Enhancement**: Improving product or service quality by analyzing customer feedback, monitoring quality metrics, and implementing corrective actions based on data-driven insights.
- **Customer Satisfaction**: Addressing customer pain points and enhancing user experience through iterative product updates, personalized services, and responsive customer support.

Innovation through Data: Data-driven innovation involves leveraging data analytics and insights to drive creative solutions, new product development, and market differentiation. Startups can innovate by:

- **Predictive Analytics**: Using predictive models to anticipate market trends, customer needs, and emerging opportunities. This enables proactive innovation and agile response to changing market dynamics.
- **Experimentation and A/B Testing**: Conducting controlled experiments and A/B tests to validate hypotheses, optimize features, and iterate on product enhancements based on empirical data and user feedback.

- **Emerging Technologies**: Exploring emerging technologies such as AI, machine learning, and IoT to unlock new business opportunities, improve operational efficiency, and deliver innovative solutions to market demands.

13 Case Study

TechCo, a fast-growing startup in the fintech sector, specializes in developing AI-powered solutions for financial institutions. With ambitions to scale rapidly, TechCo faced challenges in optimizing customer acquisition and retention strategies using data-driven decision-making.

Challenge:

- **High Customer Churn**: TechCo observed a significant churn rate among its client base, impacting revenue stability and growth projections.
- **Marketing Efficiency**: The effectiveness of marketing campaigns in targeting high-value prospects was uncertain, leading to suboptimal ROI.
- **Limited Personalization**: Lack of comprehensive customer insights hindered personalized customer interactions and proactive support initiatives.

Solution: TechCo implemented a robust data analytics framework to tackle these issues:

1. **Customer Segmentation and Churn Analysis**:
    - Utilized AI algorithms to segment customers based on transaction behavior, engagement patterns, and customer lifetime value (CLV). This segmentation allowed TechCo to identify at-risk customers early and tailor retention strategies accordingly.
2. **Predictive Analytics for Churn Prevention**:
    - Developed predictive models to forecast customer churn likelihood. By analyzing historical data and behavioral cues, TechCo proactively engaged at-risk customers with personalized offers and proactive support, reducing churn rates by 15% within six months.
3. **Marketing Optimization**:
    - Implemented A/B testing and campaign analytics to optimize marketing spend and messaging effectiveness. By analyzing conversion rates and customer response to different campaigns, TechCo achieved a 20% increase in acquisition efficiency, targeting prospects more likely to convert into long-term clients.
4. **Enhanced Customer Insights**:
    - Integrated customer feedback loops and real-time analytics to gain deeper insights into customer preferences and pain points. This enabled TechCo to

Table 2 Comparative result comparison

Results	Metrics	Achieved improvement
Customer churn reduction	Churn rate	15% decrease within 6 months
Marketing efficiency	ROI of marketing campaigns	20% increase in acquisition efficiency
Revenue growth	Annual recurring revenue (ARR)	25% increase over 1 year

iterate on product features and service offerings, enhancing overall customer satisfaction and loyalty.

Quantitative Results:

- **Churn Reduction**: Implemented strategies resulted in a 15% decrease in customer churn rate within the first six months.
- **Acquisition Efficiency**: Achieved a 20% improvement in marketing campaign ROI through targeted messaging and optimized spend allocation.
- **Revenue Growth**: Enhanced customer retention and acquisition efforts contributed to a 25% increase in annual recurring revenue (ARR) over the course of one year.

By leveraging data-driven insights, TechCo not only mitigated customer churn and optimized marketing effectiveness but also fostered a culture of continuous improvement and innovation. The strategic implementation of advanced analytics empowered TechCo to make informed decisions, drive sustainable growth, and solidify its position as a leader in the competitive fintech market.

This case study illustrates how startups can harness the power of data analytics to achieve measurable business outcomes and accelerate their journey towards long-term success in dynamic industries. The case study results presented in a tabular form in Table 2.

TechCo's strategic implementation of data-driven insights resulted in significant improvements across key metrics, including a notable reduction in customer churn, enhanced marketing efficiency, and substantial revenue growth. These outcomes demonstrate the tangible benefits of adopting data analytics to drive informed decision-making and achieve sustainable business growth in competitive markets.

14 Conclusion

The case study of TechCo exemplifies the transformative power of data-driven decision-making in driving business success. By leveraging advanced analytics to optimize customer acquisition, retention strategies, and marketing efficiency, TechCo achieved substantial improvements in key performance indicators. The implementation of predictive analytics for churn prevention not only reduced customer attrition but also fostered deeper customer insights and enhanced personalized interactions. Moreover, the strategic use of data enabled TechCo to strengthen its competitive position and accelerate revenue growth, underscoring the importance of a data-driven culture in today's dynamic business environment.

15 Future Scope

Looking ahead, the future scope for advancing data-driven strategies in startups like TechCo is promising and multifaceted:

1. **AI and Machine Learning Integration**: Further integration of AI and machine learning algorithms can enhance predictive analytics capabilities, enabling more accurate forecasting of customer behaviors and market trends.
2. **Real-time Data Analytics**: Investing in real-time data analytics infrastructure can empower startups to respond swiftly to market changes and customer feedback, facilitating agile decision-making and adaptive strategies.
3. **Personalization and Customer Experience**: Continued focus on enhancing personalization strategies through advanced segmentation techniques and dynamic content delivery can further improve customer satisfaction and loyalty.
4. **Ethical Data Use and Privacy**: Addressing ethical considerations around data use and privacy compliance will be crucial as startups scale their data-driven initiatives while maintaining customer trust and regulatory compliance.
5. **Cross-functional Collaboration**: Promoting cross-functional collaboration between data scientists, marketers, and product teams will be essential for leveraging data insights across the organization and driving innovation.

Embracing a data-driven approach not only optimizes operational efficiency and accelerates growth but also positions startups to navigate uncertainties and seize opportunities in an increasingly competitive landscape. As startups continue to evolve, the integration of data analytics will remain pivotal in shaping strategic decisions, fostering innovation, and achieving sustainable long-term success.

References

1. Joel, O.S., Oyewole, A.T., Odunaiya, O.G., Soyombo, O.T.: Navigating the digital transformation journey: strategies for startup growth and innovation in the digital era. Int. J. Manage. Entrepreneur. Res. **6**(3), 697–706 (2024)
2. Ochuba, N.A., Usman, F.O., Amoo, O.O., Okafor, E.S., Akinrinola, O.: Innovations in business models through strategic analytics and management: conceptual exploration for sustainable growth. Int. J. Manage. Entrepreneur. Res. **6**(3), 554–566 (2024)
3. Zamani, E.D., Griva, A., Spanaki, K., O'Raghallaigh, P., Sammon, D.: Making sense of business analytics in project selection and prioritisation: Insights from the start-up trenches. Inf. Technol. People **37**(2), 895–918 (2024)

4. Ibeh, C.V., Asuzu, O.F., Olorunsogo, T., Elufioye, O.A., Nduubuisi, N.L., Daraojimba, A.I.: Business analytics and decision science: a review of techniques in strategic business decision making. World J. Adv. Res. Rev. **21**(2), 1761–1769 (2024)
5. Timóteo, T.R., Cazeri, G.T., Moraes, G.H.S.M.D., Sigahi, T.F., Zanon, L.G., Rampasso, I.S., Anholon, R.: Use of AHP and grey fixed weight clustering to assess the maturity level of strategic communication management in Brazilian startups. Grey Syst.: Theory Appl. **14**(1), 69–90 (2024)
6. Martins, M.M.R.: Startup guide to AI: integrating technology for business success. Valley Int. J. Digit. Lib., 1264–1274 (2024)
7. Chaudhri, V., Pridmore, J., Mauck, C.: Assembling the start-up brand: a process framework for understanding strategic communication challenges. In: Start-up and Entrepreneurial Communication, pp. 85–100. Routledge (2024)
8. Kim, J.D.: Startup acquisitions as a hiring strategy: Turnover differences between acquired and regular hires. Strategy Sci. **9**(2), 118–134 (2024)
9. Ajalkar, D.A., Waheed, S.A., Matheen, M.A., Gupta, P., Shinde, J.P.: Futuristic business intelligence framework for start-ups. J. Autonom. Intell. **7**(2) (2024)
10. Tan, F., Zhang, Q., Mehrotra, A., Attri, R., Tiwari, H.: Unlocking venture growth: synergizing big data analytics, artificial intelligence, new product development practices, and inter-organizational digital capability. Technol. Forecast. Soc. Chang. **200**, 123174 (2024)
11. Bandini, F.: The business development of a startup and the market study of an AI innovative technology (2024)
12. Rudeloff, C., Bekmeier-Feuerhahn, S., Sikkenga, J., Barth, A.: Conditions of one-way and two-way approaches in strategic start-up communication: a qualitative comparative analysis (QCA). In: Start-up and Entrepreneurial Communication, pp. 34–58. Routledge (2024)
13. York, J.M., Turner, N., Hussels, S.: Lean startup and learning loops in entrepreneurial ventures: a systematic review. J. Knowl. Manage. Pract. **24**(1) (2024)
14. Agarwal, M., Sharma, A.: Digital marketing of startup businesses. J. Scien. Res. Technol., 8–29 (2024)
15. Grover, V., Piccoli, G.: A foundational perspective of digital business strategy for strategic leaders. In: Handbook of Research on Strategic Leadership in the Fourth Industrial Revolution, pp. 235–249. Edward Elgar Publishing (2024)
16. Kamboj, G.: Role of technology in fostering innovation and growth in startup businesses. Sustain. Dev. Goals Bus. Sustain. **37** (2024)
17. Olajiga, O.K., Olu-lawal, K.A., Usman, F.O., Ninduwezuor-Ehiobu, N.: Data analytics in energy corporations: conceptual framework for strategic business outcomes. World J. Adv. Res. Rev. **21**(3), 952–963 (2024)

Corporate Social Responsibility (CSR): A Strategic Tool for Indian Companies for Investment in the Education Sector Using a Standard Event Study

Ashu Lamba and Priti Aggarwal

Abstract This chapter examines the strategic implications of Corporate Social Responsibility (CSR) activities mandated by Indian regulations, specifically focusing on companies' investments in the education sector. With the legal requirement to devote at least 2% of average net profits to CSR initiatives, this research employs a standard event study methodology to analyze the market response to 44 unique announcements made by 38 companies in the S&P BSE 500 index between January 2011 and December 2022. The investigation delves into various forms of CSR investments, such as establishing schools and universities, supporting girls' education, and providing financial aid to underprivileged children. Using a market model approach and the BSE 30 Sensex as the market portfolio proxy, the chapter hypothesizes the absence of abnormal returns associated with the announcements of educational initiatives by Indian firms. Preliminary findings reveal nuanced market reactions, with certain days showing significant positive or negative abnormal returns. The Cumulative Average Abnormal Returns (CAAR) across diverse event windows provide a comprehensive picture of the magnitude and direction of market responses. The chapter contributes to the understanding of CSR's strategic role in the education sector and opens avenues for further research on the long-term impact of such initiatives on corporate sustainability and societal development.

Keywords Corporate Social Responsibility · Education sector · Strategic impact · Standard event study · Abnormal returns

---

A. Lamba (✉)
School of Business Studies, Vivekananda Institute of Professional Studies, New Delhi, India
e-mail: ashu.lamba@vips.edu

P. Aggarwal
Jindal Global Business School, OP Jindal Global University, Sonipat, India
e-mail: priti.aggarwal@jgu.edu.in

S. Kautish et al. (eds.), *Strategy Analytics for Business Resilience Theories and Practices*, Information Systems Engineering and Management 33,
https://doi.org/10.1007/978-3-031-82369-5_3

1 Introduction

Corporate Social Responsibility (CSR) is now recognised as an essential aspect of contemporary corporate governance, attracting significant interest from both academicians and industry practitioners [1, 2]. CSR has evolved into a fundamental pillar of business strategy and reputation management. It originates from firms taking voluntary measures to improve society outside legal requirements [1, 2, 3]. This evolution is particularly pronounced within the context of Indian companies, where indigenous philosophies and cultural values have long emphasized businesses' role as societal stewards [4].

Amidst the backdrop of increasing concerns over financial scandals and reputational risks, the importance of CSR initiatives has grown exponentially [5–7]. In India, this development was strengthened by the introduction of the Indian Companies Act in 2013, which established necessary corporate social responsibility (CSR) initiatives for becoming eligible companies [8]. Section 135 of the Act requires certain companies to apportion a portion of their profits towards CSR initiatives, reflecting the government's proactive stance in leveraging corporate resources for societal development [9].

Within the landscape of CSR, investments in the education sector hold particular significance due to their potential to drive socio-economic development in India. Education serves as a cornerstone for addressing societal disparities and fostering human capital development [8]. Therefore, understanding the strategic implications of CSR investments in education by Indian companies is paramount.

This study investigates the market response to CSR initiatives mandated by Indian regulations, specifically focusing on companies' investments in the education sector. By employing a Standard Event Study methodology, the research analyzes market reactions to CSR announcements made by Indian companies between January 2011 and December 2022. Through an examination of various forms of CSR investments in education, including the establishment of schools and universities, support for girls' education, and financial aid for underprivileged children, this study contributes to the comprehension of CSR's strategic role in promoting education and fostering long-term socio-economic development in India.

Even before the formal adoption of CSR practices, Indian companies, particularly those belonging to traditional business communities, have engaged in philanthropy and social welfare activities [10]. This statutory requirement, the first of its kind globally, underscores the Indian government's commitment to leveraging the corporate sector's resources for societal development. Furthermore, the recent amendment to the Act, shifting from a "comply or explain" to a "comply or suffer" policy, reflects the government's stringent approach towards enforcing CSR obligations [11]. In the Indian context, the education sector holds particular significance due to its critical role in fostering socio-economic development. Given the large population and socio-economic disparities in India, investments in education have the potential to yield substantial long-term benefits [8]. Therefore, it's imperative to assess the strategic implications of CSR investments in the education sector by

Indian companies. This study aims to investigate the strategic implications of CSR activities mandated by Indian regulations, with a specific focus on companies' investments in the education sector. By employing a Standard Event Study methodology, the research analyzes market responses to CSR announcements made by Indian companies between January 2011 and December 2022. The investigation will delve into various forms of CSR investments in education, including the establishment of schools and universities, support for girls' education, and financial aid for underprivileged children. Through a market model approach and utilizing the BSE 30 Sensex as the market portfolio proxy, the study hypothesizes the absence of abnormal returns associated with educational CSR initiatives by Indian firms. This study seeks to contribute to the understanding of CSR's strategic role in the education sector and open avenues for further research on the long-term impact of such initiatives on corporate sustainability and societal development. By examining market reactions to CSR announcements, the research aims to provide insights into the effectiveness of CSR as a strategic tool for Indian companies' investments in the education sector. The remaining paper is structured as: Section 2 provides a Literature Review, Sect. 3 discusses the data and methodology, Sect. 4 reports the Results, Sect. 5 offers implications, and Sect. 6 scope for future research.

2 Literature Review

Legitimacy theory, rooted in organizational theory, is pivotal in understanding corporate social responsibility (CSR) and its relationship with stakeholders. According to this theory, firms must act in ways perceived as socially responsible to maintain credibility and ensure continued success [12, 13].

Recent studies suggest that firms use sustainability disclosures to signal corporate reputation, legitimacy, and transparency, thereby gaining stakeholder support and resilience against economic downturns [14, 15]. However, amidst voluntary sustainability activities, some firms may engage in "window-dressing" strategies to manipulate corporate legitimacy, potentially leading to conflicting investor views on the genuineness of CSR engagement [16]. This highlights the complexity of CSR as it can be exploited as a short-term gimmick without delivering long-term benefits [17, 18].

3 Market Response to CSR Announcements

Legitimacy theory is a significant concept in corporate social responsibility (CSR) that investigates the relationship between firms and their stakeholders, emphasizing the significance of preserving credibility in the eyes of society. Legitimacy theory, developed primarily by organizational theorists such as Suchman [13] and Dowling and Pfeffer [12], proposes that organizations must behave in ways that the public perceives as socially responsible and acceptable in order to ensure their continued existence and success.

Previous studies suggest that adaptable firms can get backing from different stakeholder groups and lessen their vulnerability to economic recessions by using their sustainability disclosure to demonstrate their corporate reputation in the context of legitimacy and corporate transparency [14], 15.

According to Zhang et al. [16], though voluntary, few businesses may utilize sustainability initiatives to use corporate legitimacy, in light of the extreme pressures experienced during the epidemic. As a result, investors may hold divergent opinions about a company's participation in sustainability initiatives. CSR is often exploited as a gimmick without delivering a long-term benefit [17, 18].

Previous research has extensively analyzed the market response to CSR announcements, shedding light on various aspects of investor sentiment and market dynamics. Studies that examined the effects of CSR rating announcements on European markets found that trading volume fluctuates before and after CSR rating announcements, with an upgrade in CSR rating leading to a significant decline in Credit Default Swap (CDS) spreads for the sample firms [19, 20]. In the Indian setting, research has shown that mandated CSR adoption can influence firm value, with differential effects observed for foreign and domestic firms [21, 22]. Furthermore, CSR-related news events in Spain and Austria have demonstrated significant abnormal returns and market reactions, highlighting the importance of CSR activities for investor perception and firm performance [23–25]. Additionally, studies explore the impact of CSR reputation on stock performance, showing that failures to report product defects can be counteracted by positive CSR reputation effects [26].

4 CSR and Financial Performance

CSR is a crucial strategy for building strong stakeholder relationships. Firms' social commitment fosters credibility and reciprocity among stakeholders, leading to increased market opportunities and lower transaction costs [27, 28]. Mixed results have been reported from studies on the relationship between CSR and corporate financial performance (CFP).

While few studies affirm a positive association between CSR and CFP [29–31, 32], others have reported negative [33] or neutral relationships [34, 35, 36, 37]. Limited research exists on the impact of CSR expenditure on financial performance, necessitating further investigation to understand the implications of mandatory CSR provisions introduced by the Indian Companies Act of 2013 [8, 38].

5 Education and CSR

Education is widely recognized as a critical determinant of economic and human development, with numerous studies highlighting its positive impact on productivity, individual salaries, and GDP growth [7, 39, 40]. Consequently, businesses often invest in education as part of their CSR initiatives, recognizing its role in fostering community engagement and long-term socio-economic development [41]. Given the strategic importance of education in driving societal progress, understanding the implications of CSR investments in the education sector is necessary for both businesses and policymakers.

6 Data and Methodology

The study focuses on analyzing education-related Corporate Social Responsibility (CSR) announcements made by the top 500 Indian companies across various industries from 2011 to 2022. This period was selected to capture a comprehensive dataset reflecting CSR initiatives in the education sector within the Indian corporate landscape. To collect the necessary data systematically, we employed a search strategy utilizing Google to identify news articles and press releases containing CSR announcements related to investments in education. The search queries included company names combined with keywords such as "education contribution" or "CSR initiative in education." This method ensured the retrieval of relevant and reliable information regarding CSR activities undertaken by Indian firms across diverse sectors.

The final sample comprises 44 education-related CSR announcements made by 38 Indian companies, representing various industries and sectors in the time period from January 2011 to December 2022. These announcements encompass contributions and initiatives aimed at promoting education and skill development across different regions of India.

To analyze the market response to CSR announcements, a standard event study methodology was employed [42]. This methodology allows for the examination of abnormal stock returns surrounding the timing of CSR announcements, providing insights into investor sentiment and market dynamics in response to CSR initiatives. The news about the contribution to education is examined on the first day of the announcement to see how it could affect the stock returns of the sample companies.

Abnormal returns ($\mathrm{AR_{it}}$) were calculated using the following equation:

$$\mathrm{AR_{it}} = \mathrm{R_{it}} - (\alpha \mathrm{i} + \beta_{\mathrm{i}} \mathrm{R_{mt}}) \tag{1}$$

where

AR_{it} is the abnormal return of the firm 'i' on day 't; R_{it} is the log return of the security's price for a firm 'i' on the day t; R_{mt} is the rate of return on the benchmark index on day t and α_i and β_i are OLS parameters found from the estimation period.

This equation enables the identification of abnormal stock price movements associated with CSR announcements, adjusted for market-wide trends captured by the benchmark index.

The estimation window spans from − 261 to − 1 days from the event day which facilitated the calculation of estimates alpha and beta. Stock price data for both the estimation and event windows were sourced from the CMIE PROWESS database, ensuring consistency and accuracy in the analysis.

Next, we sum the average abnormal returns, called Cumulative Average Abnormal Returns (CAAR), for different windows in the event window to examine the security's performance over a defined period. Further t-statistics is calculated to check the significance of AAR and CAAR, respectively.

7 Results

The event study results are shown in Table 1, which offers insights into how the market responded to Indian businesses' announcements of CSR initiatives linked to education. It includes average abnormal returns (AAR), and associated t-statistics for various event windows, facilitating the interpretation of the findings and their statistical significance.

On the day of the event (day 0), the AAR is 0.19%, which is not statistically significant. However, statistically significant positive market reactions are observed on t − 7, while significant negative reactions are noted on t − 1, t + 1, and t + 9. However, there were no significant reactions on the announcement day itself (Day 0) or during the remaining post-event days. These findings suggest that the market initially viewed the CSR initiatives positively but subsequently adjusted its perception in subsequent days. The observed negative average abnormal returns indicate that, on average, the market did not react positively to CSR initiatives in education. This contrasts with the study's hypotheses, suggesting potential disparities between theoretical expectations and empirical outcomes. Other factors, such as broader market conditions, news flow, or specific events happening concurrently, may have influenced the market's reactions as well. The analysis of market reactions reveals intriguing insights into investor sentiment and the perceived value of CSR activities in the education sector. Notably, significant variations in abnormal returns are observed across different event windows, shedding light on the nuanced dynamics at play as shown in Fig. 1.

Table 2 illustrates the CAAR and t-statistics for different event windows, delineating the market's reaction before, during, and after education-related CSR announcements by Indian companies.

In the CAAR (− 10, − 1) window, there is a small positive CAAR of 0.25%, although statistically insignificant with a t-statistic of 0.38. Similarly, CAAR (− 5,

Table 1 Average abnormal returns of firms investing in education sector

Days	AAR (%)	t_stats_{AAR}
t − 10	0.02	0.06
t − 9	− 0.14	− 0.55
t − 8	− 0.23	− 0.89
t − 7	0.46**	2.13
t − 6	0.03	0.10
t − 5	− 0.31	− 1.77
t − 4	0.27	1.22
t − 3	− 0.05	− 0.22
t − 2	0.20	0.86
t − 1	− 0.37**	− 2.21
0	0.19	0.70
t + 1	− 0.48**	− 2.11
t + 2	− 0.09	− 0.33
t + 3	0.10	0.30
t + 4	0.11	0.44
t + 5	0.14	0.66
t + 6	0.09	0.38
t + 7	0.39	1.61
t + 8	0.06	0.25
t + 9	− 0.67***	− 3.07
t + 10	0.21	0.91

Notes ***means Significant at 1% level, **means Significant at 5% level and *means Significant at 10% level

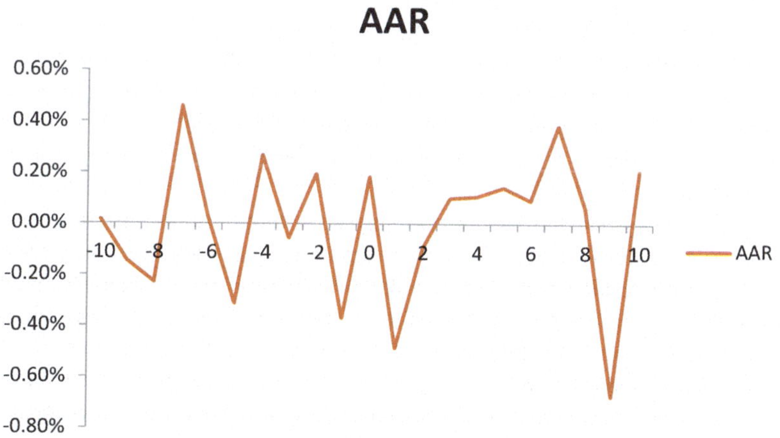

Fig. 1 Average abnormal returns of the firms announcing investment in education sector

Table 2 Cumulative average abnormal returns of the sample firms during various intervals

Intervals	Event windows	CAAR (%)	t_statsCAAR
Before event	CAAR(– 10, – 1)	0.25	0.38
	CAAR(– 5, – 1)	0.14	0.27
	CAAR(– 3, – 1)	0.42*	1.70
During event	CAAR(– 10, 10)	– 0.27	– 0.25
	CAAR(– 5, 5)	– 0.52	– 0.70
	CAAR(– 3, 3)	– 0.34	– 0.92
	CAAR(– 1, 1)	0.02	0.05
After event	CAAR(0, 10)	– 0.52	– 0.68
	CAAR(0, 5)	– 0.54	– 0.86
	CAAR(0, 3)	– 0.76	– 1.28
	CAAR(0, 1)	– 0.18	– 0.41

Notes ***Significant at 1% level, **Significant at 5% level, *Significant at 10% level

– 1) shows a slightly positive cumulative abnormal return of 0.14%, also statistically insignificant (t-statistic = 0.27). However, CAAR (– 3, – 1) exhibits a more substantial positive cumulative abnormal return of 0.42%, which is statistically significant at the 5% level (t-statistic = 1.70), indicating a slightly positive market reaction leading up to the CSR announcements.

The CAAR (– 10, 10) window displays returns of – 0.27%, although statistically insignificant (t-statistic = – 0.25). A similar pattern is observed in CAAR (– 5, 5) and CAAR (– 3, 3), with negative abnormal returns of -0.52% and -0.34%, respectively, both statistically insignificant. In the CAAR (– 1, 1) window, the cumulative abnormal return is close to zero at 0.02%, indicating no significant deviation from expected returns during the event period.

Following the event, CAAR (0, 10) shows a negative cumulative abnormal return of – 0.52%, statistically insignificant (t-statistic = – 0.68). Similarly, CAAR (0, 5) and CAAR (0, 3) demonstrate negative returns of – 0.54% and – 0.76%, respectively, with statistically insignificant t-statistics. However, in the CAAR (0, 1) window, the cumulative abnormal return is relatively smaller at – 0.18%, also statistically insignificant.

These results suggest that while there may be some positive market reactions leading up to education-related CSR announcements, particularly in the days closer to the event, the overall impact on cumulative abnormal returns remains negative. The statistically insignificant t-statistics across various event windows indicate no conclusive evidence of abnormal market reactions during those periods.

This is evident from the negative cumulative returns observed in various event windows, such as CAAR (0, 10), CAAR (0, 5), and CAAR (– 5, 5). This suggests that, on average, the market did not react positively to the CSR initiative in education. The pre-event window CAAR (– 10, – 1) shows a small positive cumulative abnormal return, indicating a slightly positive market reaction leading up to the CSR announcement. However, this positive impact is not statistically significant. In the post-event windows, no statistically significant abnormal returns were observed, indicating that the market did not exhibit a significant reaction to the CSR announcement immediately after the days following the event.

8 Implications

Based on the finding of no significant results in the study examining the impact of news about companies investing in education-related activities on stock returns, several implications can be drawn. The study's findings suggest that investors may not perceive education-related CSR initiatives as materially impactful in the short term. This implies that companies may need to explore alternative strategies or initiatives to generate stronger market responses aligned with investor expectations. Effective communication about education-related CSR initiatives remains crucial for maintaining stakeholder trust and enhancing corporate reputation. Transparency and clarity in communication strategies can convey the long-term value and societal impact of these initiatives, despite the lack of immediate market effects. Although short-term market reactions may be limited, companies should adopt a long-term perspective when evaluating the impact of CSR initiatives in education. Recognizing the potential long-term benefits, such as enhanced employee skills and community development, is essential for sustaining these initiatives. Engaging with diverse stakeholders beyond investors, including employees, customers, communities, and regulators, is critical for assessing the broader societal impact of education-related CSR initiatives. Tailoring programs to address specific stakeholder needs and concerns can enhance relevance and impact. Integration with Business Strategy: Companies should strategically integrate education-related CSR initiatives into their broader business strategy to maximize effectiveness and alignment with organizational goals. Creating synergies between CSR and core business activities can benefit both the community and the business, enhancing brand reputation and fostering innovation. Given the dynamic nature of the business environment and societal needs, continuous evaluation and adaptation of CSR initiatives are essential. Regular monitoring, measurement, and reporting of outcomes can demonstrate accountability and drive continuous improvement in CSR performance.

Overall, while the immediate market reaction to news about education investments may be limited, companies can still derive value from these initiatives by focusing on long-term sustainability, stakeholder engagement, and strategic integration with business objectives. By taking a holistic approach to CSR and education initiatives, companies can contribute meaningfully to societal development while also enhancing their competitive advantage and reputation in the long run. Integrating Corporate Social Responsibility (CSR) initiatives in education into a company's overall strategy requires careful planning, alignment with business objectives, and a commitment to long-term impact.

9 Scope of Future Research

Although this study provides relevant information about how the market reacts to CSR (corporate social responsibility) initiatives in the education sector, several limitations must be acknowledged. Focusing solely on Indian companies may restrict the applicability of findings to other geographical regions. Including companies from diverse locations in future studies would enable cross-country comparisons of CSR practices and their impact on the education sector. The study's sample comprised announcements from companies across various industries, potentially overlooking variations in CSR practices among different sectors. Exploring CSR initiatives across industries could provide a more comprehensive understanding of their implications for education. While the study analyzed market reactions to CSR announcements, it did not directly assess the long-term financial implications of these initiatives. Future research could employ financial performance metrics to evaluate the economic impact of CSR investments in education. The event study methodology used in this research focused on analyzing market reactions to CSR announcements on specific dates. Future studies could employ complementary methods, such as qualitative case studies or longitudinal analyses, to provide a more holistic overview of CSR's role in promoting education. The study relied on publicly available sources for data collection, which may vary in accuracy and reliability. Future research could utilize more robust data sources, such as company reports or regulatory filings, to ensure data quality and reliability.

References

1. Carroll, A.B.: The pyramid of corporate social responsibility: toward the moral management of organizational stakeholders. Bus. Horiz.Horiz. **34**(4), 39–48 (1991)
2. Walton, C.C.: Corporate Social Responsibilities. Wadsworth, Belmont, CA (1967)
3. Wang, Y.-S., Chen, Y.-J.: Corporate social responsibility and financial performance: event study cases. J. Econ. Interac. Coord.Interac. Coord. **12**(2), 193–219 (2017). https://doi.org/10.1007/s11403-015-0161-9
4. Baxi, C.V., Prasad, A.: Corporate Social Responsibility: Concepts and Cases. Excel Books, New Delhi (2005)
5. Copeland, T.: Valuation and Measuring the Value of Companies. Wiley & Sons, New York (1994)
6. Davis, K.: Can business afford to ignore corporate social responsibilities. Calif. Manage. Rev. **2**, 70–76 (1960)
7. Friedman, M. (1970). The Social Responsibility of Business is to Increase its Profits, pp. 122–126. New York Times Magazine
8. Agrawal, A., Sahasranamam, S.: Corporate social entrepreneurship in India. South Asian J. Glob. Bus. Res. **5**(2), 214–233 (2016). https://doi.org/10.1108/sajgbr-12-2014-0098
9. Kapoor, G.K., Dhamija, S.: Mandatory CSR spending—Indian experience. Emerg. Econ. Stud. **3**(1), 98–112 (2017). https://doi.org/10.1177/2394901517696645
10. Sood, A., Arora, B.: The political economy of corporate responsibility in India, United Nations Research Institute for Social Development (UNRISD) Programme Paper (2006)
11. Ministry of Corporate Affairs: Report of the high level committee on CSR 2018 (2019). Available at: https://www.mca.gov.in/Ministry/pdf/CSRHLC_13092019.pdf

12. Dowling, J., Pfeffer, J.: Organizational legitimacy: social values and organizational behavior. Pac. Sociol. Rev.Sociol. Rev. **18**(1), 122–136 (1975). https://doi.org/10.2307/1388226
13. Suchman, M.C.: Managing legitimacy: strategic and institutional approaches. Acad. Manag. Rev.Manag. Rev. **20**(3), 571–610 (1995). https://doi.org/10.2307/258788
14. Ding, W., Levine, R., Lin, C., Xie, W.: Corporate immunity to the COVID-19 pandemic. J. Financ. Econ.Financ. Econ. **141**(2), 802–830 (2021)
15. Zhang, L., Shan, Y.G., Chang, M.: Can CSR disclosure protect firm reputation during financial restatements? J. Bus. Ethics **173**, 157–184 (2021)
16. Zhang, J., Zheng, C., Shan, Y.G.: What accounts for the effect of sustainability engagement on stock price crash risk during the COVID-19 pandemic—agency theory or legitimacy theory? Int. Rev. Fin. Anal., 103167 (2024)
17. Laufer, W.S.: Social accounting and corporate greenwashing. J. Bus. Ethics **43**(3), 253–261 (2003). https://doi.org/10.1023/A:1022962719299
18. Ramus, C.A., Montiel, I.: When are corporate environmental policies a form of greenwashing? Bus. Soc. **44**(4), 377–414 (2005). https://doi.org/10.1177/0007650305278120
19. Cellier, A., Chollet, P., Gajewski, J.-F.: Do investors trade around social rating announcements? Eur. Financ. Manag.Financ. Manag. **22**(3), 484–515 (2016). https://doi.org/10.1111/eufm.12066
20. Drago, D., Carnevale, C., Gallo, R.: Do corporate social responsibility ratings affect credit default swap spreads? Corp. Soc. Responsib. Environ. Manag.Responsib. Environ. Manag. **26**(3), 644–652 (2019). https://doi.org/10.1002/csr.1709
21. Manchiraju, H., Rajgopal, S.: Does Corporate Social Responsibility (CSR) create shareholder value? Evidence from the Indian Companies Act 2013. J. Account. Res. **55**(5), 1257–1300 (2017). https://doi.org/10.1111/1475-679X.12174
22. Panwar, R., Pandey, V., Suddaby, R., Vidal, N.G.: Did India's CSR mandate enhance or diminish firm value? Bus. Soc. **62**(2), 401–433 (2023). https://doi.org/10.1177/00076503221085962
23. Ender, M., Brinckmann, F.: Impact of CSR-Relevant news on stock prices of companies listed in the Austrian traded index (atx). Int. J. Fin. Stud. **7**(3), 36 (2019). https://doi.org/10.3390/ijfs7030036
24. Pérez, A., García de los Salmones, M.d., López-Gutiérrez, C.: Market reactions to CSR news in different industries. Corp. Commun. **25**(2), 243–261 (2020). https://doi.org/10.1108/CCIJ-05-2019-0056
25. Pérez, A., López-Gutiérrez, C., García de los Salmones, M.d.: Do all CSR news affect market value equally? Social Responsib. J. **16**(8), 1107–1123 (2020). https://doi.org/10.1108/SRJ-03-2019-0116
26. Afrin, R., Peng, N., Bowen, F.: The wealth effect of corporate water actions: how past corporate responsibility and irresponsibility influence stock market reactions. J. Bus. Ethics **180**(1), 105–124 (2022). https://doi.org/10.1007/s10551-021-04881-6
27. Barnett, M.L.: Stakeholder influence capacity and the variability of financial returns to corporate social responsibility. Acad. Manag. Rev.Manag. Rev. **32**(3), 794–816 (2007)
28. Barnett, M.L., Salomon, R.M.: Does it pay to be really good? Addressing the shape of the relationship between social and financial performance. Strateg. Manag. J.. Manag. J. **33**(11), 1304–1320 (2012)
29. Garay, L., Font, X.: Doing good to do well? Corporate social responsibility reasons, practices and impacts in small and medium accommodation enterprises. Int. J. Hosp. Manag.Manag. **31**(2), 329–337 (2012)
30. Nguyen, C.T., Nguyen, L.T., Nguyen, N.Q.: Corporate social responsibility and financial performance: the case in Vietnam. Cogent Econ. Fin. **10**(1), 2075600 (2022)
31. Rodríguez, F.J.G., Cruz, Y.D.M.A.: Relation between social-environmental responsibility and performance in hotel firms. Int. J. Hosp. Manag.Manag. **26**(4), 824–839 (2007)
32. Waddock, S.A., Graves, S.B.: The corporate social performance-financial performance link. Strateg. Manag. J.. Manag. J. **18**(4), 303–319 (1997)
33. El Ghoul, S., Guedhami, O., Kim, Y.: Country-level institutions, firm value, and the role of corporate social responsibility initiatives. J. Int. Bus. Stud. **48**(3), 360–385 (2017). https://doi.org/10.1057/jibs.2016.4

34. Gollop, F., Roberts, M.J.: Environmental regulation and productivity growth: The case of fossil-fuel electric power generation. J. Polit. Econ. **91**(4), 654–674 (1983)
35. Aupperle, K.E., Carroll, A.B., Hatfield, J.D.: An empirical examination of the relationship between corporate social responsibility and profitability. Acad. Manag. J.Manag. J. **28**(2), 446–463 (1985)
36. Makni, R., Francoeur, C., Bellavance, F.: Causality between corporate social performance and financial performance: evidence from Canadian firms. J. Bus. Ethics **89**(3), 409–422 (2009). https://doi.org/10.1007/s10551-008-0007-7
37. Peng, C.W., Yang, M.L.: The effect of corporate social performance on financial performance: the moderating effect of ownership concentration. J. Bus. Ethics. **123**(1), 171–182 (2014). https://doi.org/10.1007/s10551-013-1809-9
38. Margolis, J.D., Hillary, E.A., & Walsh, J.R. (2007). Does it pay to be good? A meta-analysis and redirection of research on the relationship between corporate and social and financial performance. https://doi.org/10.2139/ssrn.1866371
39. Heckman, J.J., Moon, S.H., Pinto, R., Savelyev, P.A., Yavitz, A.: The rate of return to the HighScope Perry preschool program. J. Public Econ. **94**(1–2), 114–128 (2010)
40. Patrinos, H.A.: Estimating the return to schooling using the Mincer equation. IZA World of Labor (2016)
41. Aguinis, H., Glavas, A.: What we know and don't know about corporate social responsibility: a review and research agenda. J. Manag.Manag. **38**(4), 932–968 (2012)
42. Brown, S.J., Warner, J.B.: Using daily stock returns: the case of event studies. J. Financ. Econ.Financ. Econ. **14**(1), 3–31 (1985)
43. Breton, T.R.: The role of education in economic growth: theory, history and current returns. Educational Research **55**(2), 121–138 (2013)
44. Gao, N., & Qayyum, A. (2022). The evolution of firms' social responsibility and stock performance. Social Responsibility Journal. Advance online publication. https://doi.org/10.1108/SRJ-06-2021-0222
45. Groening, C., Kanuri, V.K.: Investor reactions to concurrent positive and negative stakeholder news. J. Bus. Ethics **149**(4), 833–856 (2018). https://doi.org/10.1007/s10551-016-3065-2
46. Hanushek, E.A.: Will more higher education improve economic growth? Oxf. Rev. Econ. Policy. Rev. Econ. Policy **32**, 538–552 (2016)
47. Luo, X., Bhattacharya, C.B.: The debate over doing good: corporate social performance, strategic marketing levers, and firm-Idiosyncratic risk. J. Mark. **73**(6), 198–213 (2009). https://doi.org/10.1509/jmkg.73.6.198
48. McGuire, J.B., Sundgren, A., Schneeweis, T.: Corporate social responsibility and firm financial performance. Acad. Manag. J.Manag. J. **31**(4), 854–872 (1988)
49. McWilliams, A., Siegel, D.: Corporate social responsibility and financial performance: correlation or misspecification? Strateg. Manag. J.. Manag. J. **21**(5), 603–609 (2000)
50. McWilliams, A., Siegel, D.: Corporate social responsibility: a theory of the firm perspective. Acad. Manag. Rev.Manag. Rev. **26**(1), 117–127 (2001)
51. O'Donovan, G.: Environmental disclosures in the annual report: extending the applicability and predictive power of legitimacy theory. Acc. Audit Acc. J. **15**(3), 344–371 (2002). https://doi.org/10.1108/09513570210435870
52. Velayutham, E., Ratnam, V.: Corporate social responsibility, business group affiliation and shareholder wealth: evidence from an emerging market. Social Responsib. J. **18**(4), 723–743 (2022). https://doi.org/10.1108/SRJ-08-2020-0337
53. World Business Council for Sustainable Development: Corporate Social Responsibility: Meeting Changing Expectations. World Business Council for Sustainable Development, Geneva (1999)

Strategic Analytics for Synergizing Digital and Real Economies

Guo Yantong, Hayyan Nassar Waked, S. B. Goyal, and Deepmala Singh

Abstract The article aims to explore the status quo and influencing factors of the Synergy development of digital economy and real economy in Jiaozuo City, Henan, China. This study collects data from Jiaozuo City from 2011 to 2022, and then uses the entropy value method, the coupling coordination degree model, and the gray correlation degree to measure the level of development of the Synergy of the digital economy and the real economy in Jiaozuo City, aiming to explore the development of the digital-real economics synergy in Jiaozuo City in the past 12 years and the results show that Jiaozuo City's digital economy and real economy have developed from the initial severe disorder in 2011 to the high-quality coordination in 2022 and the synergistic promotion effect is presented, but the development level of the real economy has shown a trend of gradually lagging behind the development level of the digital economy. Moreover, the influence factors of the digital and real economics synergy in Jiaozuo City are selected by Economic Development (ED), Openness (OP), Economic Efficiency (EE), R&D development (R&D) and Industrial Structure (IS), which are explored by using grey correlation analysis to verify the key influence factors of digital-real economics synergy in Jiaozuo City based on the analysis of the empirical study. R&D is the most important influencing factor, followed by ED, OP, IS and EE; This study strongly recommends that policymakers in Jiaozuo City consider incentivizing enterprises to increase R&D investment, set up an R&D team led by high-level talents, and build an R&D platform to promote the long-term development of the digital-real economics synergy.

Keywords Digital economy · Real economy · Digital-real economics synergy · Jiaozuo city

G. Yantong · H. N. Waked (✉) · S. B. Goyal (✉)
City Graduate School, City University, 46100 Petaling Jaya, Malaysia
e-mail: dr.hayyan@city.edu.my

S. B. Goyal
e-mail: drsbgoyal@gmail.com

D. Singh
Symbiosis Centre for Management Studies (SCMS), Symbiosis International (Deemed University) (SIU), Mouza-Wathoda, Nagpur, Maharashtra, India

S. Kautish et al. (eds.), *Strategy Analytics for Business Resilience Theories and Practices*, Information Systems Engineering and Management 33,
https://doi.org/10.1007/978-3-031-82369-5_4

1 Introduction

The Digital Economy (DE) refers to an economic system where the primary factors of production and service delivery are digital data and technologies, encompassing a wide range of activities from online services and e-commerce to financial technologies and digital platforms. It is characterized by the use of digital data in binary form to enhance efficiency, productivity, and value creation across various sectors, including agriculture, construction, and basic industries, not just the information and telecommunications sector [5]. The Real Economy (RE) encompasses the production, distribution, and consumption of tangible goods and services, forming the backbone of economic activities that directly impact people's lives. It includes sectors such as manufacturing, agriculture, and services, which are essential for creating real wealth, ensuring full employment, and promoting economic growth [8].

Digital-real economics synergy represents a transformative process where digital technologies and data-driven innovations are seamlessly embedded into traditional economic sectors to enhance efficiency, productivity, and sustainability. This synergy is pivotal for fostering high-quality economic development and is seen as a crucial strategy for modernizing industrial systems and enhancing national competitiveness [27]. Jiaozuo, located in the northwestern part of China's Henan Province, is a city that thrives on energy, however, over time, extensive mining activities have led to land subsidence problems and challenges with environmental issues. In order to promote the city's sustainable development, the city has also been working on the digital transformation of the manufacturing industry to meet the requirements of the digital age.

The digital economy is an important engine for China to realize high-quality development. According to the *Digital China Development Report (2022)*, the scale of China's digital economy has reached 50.2 trillion yuan in 2022, and the proportion of gross domestic product (GDP) has risen to 41.5%. In January 2022, the State Council issued the *14th Five-Year Plan for the Development of the Digital Economy*, which takes data as the main direction of future economic development and the strong economic correlation between digital technology and the real economy as the main means of realizing the modernization and transformation of traditional industries (National Internet Information [32]). In October of the same year, General Secretary Xi Jinping re-emphasized in the report of the 20th Party Congress that accelerate the development of the digital economy and promote the deep synergy of the digital economy with the real economy. The integrated development of digital economy and real economy has become an important grip and impetus for promoting China's economic growth. Therefore, studying and promoting the synergy of digital and real development is of great theoretical and practical significance for China to realize the goal of high-quality development.

Jiaozuo City, as an industrial town in Henan Province, has demonstrated positive exploration and practice in this synergy process. Jiaozuo city has made significant progress in the construction of digital infrastructure. According to the *14th Five-Year Plan for the Development of Digital Economy in Jiaozuo City*, the city has achieved

5G network coverage in major urban areas and built a number of big data centers and cloud computing platforms. These infrastructures have laid a solid technical foundation for the development of the digital economy [19] Meanwhile, Jiaozuo city has made some achievements in digital infrastructure construction, the overall level still needs to be improved. According to the *Peking University Digital Inclusive Finance Index*, the level of digital infrastructure construction in Jiaozuo is 125.36, lower than the Henan province average 340.81 on 2020 [16]. Especially in rural and remote areas, network coverage and speed are still unsatisfactory, limiting the further development of the digital economy. This suggests that Jiaozuo City still needs to increase investment and planning in digital infrastructure development to ensure a balanced development across the city. Meanwhile, Jiaozuo City faces challenges in the introduction and cultivation of high-end digital talents. Due to the lack of policies and environment to attract high-level talents, many enterprises have difficulties in meeting the demand for talents in areas such as intelligent manufacturing and data analysis. According to the latest statistics, the shortage of high-end digital talents in Jiaozuo has reached 20 percent, Jiaozuo absorbed 2036 master's and PhD high-level talents to meet this demand [18]. However, Jiaozuo City's policy support in promoting the synergy of the digital and real economy needs to be continuously optimized. Existing policies are mostly focused on infrastructure construction and technology promotion, and lack specific support measures for the actual needs of enterprises [20, 21]. In addition, there are problems such as information asymmetry and lack of implementation in the process of policy implementation, which have affected the effectiveness of the policies [22]. Therefore, It is necessary to study on the deep synergy of Digital Economy and Real Economy in Jiaozuo City.

2 Literature Review

2.1 Digital Economy

The digital economy in Henan, China, has had a substantial impact on several sectors, promoting economic development of high quality and optimizing the structure. The digital economy has had a significant effect on the logistics business in Henan. The progress in digital technologies is closely related to the economic growth in the tertiary sector, which includes logistics. This correlation has been proved by a regression model [33]. Henan's economic growth has been swift and consistent since the implementation of reforms and opening up policies. Through resolving information gaps and facilitating the connection between small-scale farmers and larger markets, the digital economy significantly contributes to the eradication of poverty. Digital platforms that improve market accessibility and offer vital data, such pricing and demand patterns, are what enable this connectedness and help farmers make well-informed decisions.

Therefore, the digital economy can improve the effectiveness and accessibility of the agricultural economy [38]. The digital economy significantly enhances innovation and resource allocation efficiency in the manufacturing sector, contributing to high-quality development [36]. These findings emphasize the diverse and important role of the digital economy in promoting Henan's economic growth. They also show the necessity of ongoing investment in digital infrastructure, fostering talent, and implementing supportive policies to fully exploit its potential.

2.2 Real Economy

Jiaozuo, located in Henan Province, China, exhibits a diverse economic environment marked by industrial growth, ecological difficulties, and changing socio-economic factors. The city's real economy has traditionally been propelled by its industrial sectors, namely mining and manufacturing. The Mining and Processing of Ferrous Metal Ores is the industry sector that has the highest energy eco-efficiency. On the other hand, the Manufacture of Special Purpose Machinery is the industry sector that leads in solid waste-related eco-efficiency. This indicates a strong industrial foundation with different environmental implications [6].

The development of the real economy is usually calculated by the value added of all industries other than the financial and real estate industries [26], the growth rate of the real economy in Jiaozuo City continued to decline after 2010, the downward pressure is huge, resulting in such a phenomenon is mainly for the city of Jiaozuo City, as a typical resource-exhausted city, the slow economic transformation, science and technology innovation capability is insufficient, coupled with the impact of the COVID19, so that the original lack of growth in the real. The economy is further aggravated, and the growth rate of the real economy of Jiaozuo City in 2020 was even – 20.4%, and then gradually review as the outbreak passed [14]. This expansion of industry has incurred environmental consequences.

The Environmental Kuznets Curve analysis of Jiaozuo reveals a U-shaped pattern, suggesting that although the pollution per capita GDP is generally decreasing, the city is currently undergoing industrial expansion that has substantial environmental consequences [41]. Therefore, there is an urgent need to transform the economic development mode of Jiaozuo City and accelerate of integrating the digital economy with the real economy in order to promote sustainable economic development.

2.3 Synergy of Digital Economy and Real Economy

Empirical studies conducted provide valuable insights into the economic transformation and development of the region by examining the merger of the digital economy and the real economy. Research has indicated that the digital economy has a substantial role in promoting the growth and improvement of many industries, such as agriculture, manufacturing, and services. This is achieved through the facilitation of comprehensive digital transformation and the enhancement of fundamental competitiveness [25].

The logistics business in Henan has experienced significant advantages from digital improvements. Regression models have shown a association between digitalization and economic growth in the tertiary sector [33]. The incorporation of digital technologies such as big data, cloud computing, and artificial intelligence has played a crucial role in improving the allocation of resources and facilitating the development of innovative retail and industrial models [15].

An empirical investigation conducted in Kaifeng City, which is located in Henan, utilized models such as the Super-SBM and Tobit models. The findings indicate that the digital economy has a substantial positive impact on the manufacturing sector. However, the extent of this impact differs depending on the business type, with capital-intensive industries experiencing the greatest benefits [35]. In addition, the spatial Durbin model emphasizes the contribution of the digital economy in decreasing carbon emissions in agriculture, namely in major grain-producing areas and western regions. This suggests a spatial spillover effect that may be significant for Henan's agricultural industry [37].

The research methods of the existing empirical studies related to the development of digital and real economy synergy are mainly based on multiple linear regression, Super-SBM and Tobit models, and there is a lack of analysis from the perspectives of coordination and correlation; at the same time, the research areas of digital-real synergy are mainly concentrated at the national and provincial levels, and there are very few analyses at the city level, and there is no corresponding analysis of the development of the digital economy and real economy of Jiaozuo City, Henan Province.

Based on the above literature, this study aims to empirically analyze the current situation of the synergy of digital economy and real economy in Jiaozuo City. The study adopts the entropy value method and the coupling coordination degree model to measure the development level of digital-real synergy in Jiaozuo and clarifies its current development situation. In addition, this paper also explores the key factors affecting the synergy of digital and real economy by using gray correlation analysis and clarifies the specific path of digital-real synergy in Jiaozuo City, with a view to providing a reference basis for its digital and real economy synergy development.

3 Methodology

3.1 Indicator Selection

The Statistical Classification of the Digital Economy and its Core Industries 2021 defines the digital economy as a series of economic activities in which data resources are the key factors of production, modern information networks are the key carriers,

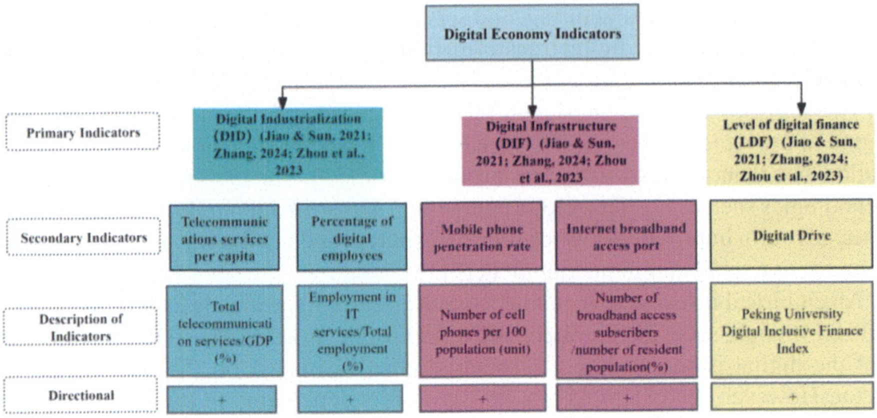

Fig. 1 Jiaozuo city digital economy evaluation indicator system

and effective use of information and communication technologies is an important driving force for improving efficiency and optimizing economic structure [1], and divides digital economy into digital industrialization and industrial digitization, of which digital industrialization is the core industry of digital economy, including digital product manufacturing, digital product service industry, digital technology application industry, and digital factor-driven industry, and industrial digitization includes digital efficiency improvement industry. Based on this definition and classification, and considering the availability of data, this paper establishes three primary indicators and five secondary indicators for digital industrialization, digital infrastructure, and digital financial level [17, 42, 44], specifically as shown in Fig. 1.

This study draws on the construction of evaluation index for real economy development by previous researchers, and selects 10 secondary indexes from the three dimensions of industrial scale, industrial structure, and institutional culture to construct the evaluation index system for real economy development in Jiaozuo City [9, 12] as shown in Fig. 2.

3.2 Measurement Methods and Data Sources

3.2.1 Entropy Method

The entropy method plays a crucial role in economic forecasting by characterizing the complexity of systems and analyzing probability distributions to enhance predictive models. It has been utilized in various fields, including financial time series forecasting [10].

This paper uses SPSSAU software and utilizes the entropy value method to comprehensively measure the development status of digital economy and real

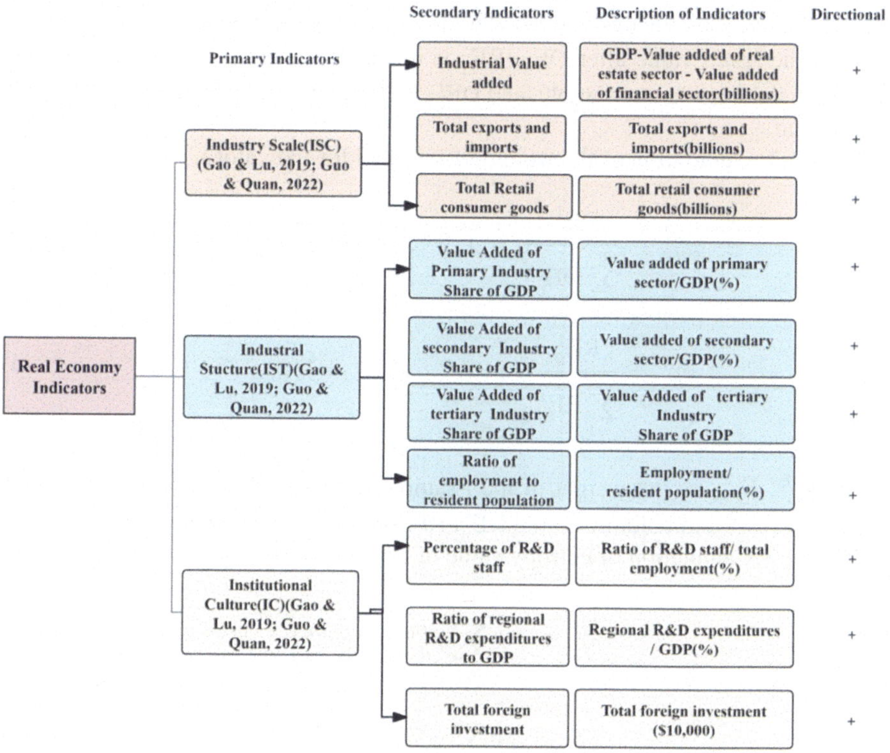

Fig. 2 Jiaozuo city real economy evaluation indicator system

economy in Jiaozuo City. The principle of entropy value method is to calculate the information entropy of a particular indicator, the entropy value can evaluate the degree of dispersion of the indicator, the greater the degree of dispersion indicates that the indicator has more influence on the comprehensive evaluation score to get the weight of the indicator, and then get the comprehensive score of the indicator system, which can be used as a comprehensive index for evaluating the state of development of the economy. The steps are as follows:

Phases 1. Since the indicators have a positive or negative influence on the system, it is necessary to carry out dimensionless processing while retaining the indicator data with central dispersion. After centering the data according to its minimum value, and then scaled by the extreme deviation, the data are rotated by the minimum value of one unit, and are bounded to between [0, 1]. The formula is as follows:

$$\mathrm{DE}_{\mathrm{ij}}^{1} = \frac{\mathrm{DE}_{ij} - \min(\mathrm{DE}_{ij})}{\max(\mathrm{DE}_{ij}) - \min(\mathrm{DE}_{ij})} \tag{1}$$

$$\mathrm{RE}_{\mathrm{ij}}^{1} = \frac{\mathrm{RE}_{ij} - \min(\mathrm{RE}_{ij})}{\max(\mathrm{RE}_{ij}) - \min(\mathrm{RE}_{ij})} \tag{2}$$

where DE_{ij}, RE_{ij} are the updated value for period i under secondary indicator j, DE_{ij}^{1}, RE_{ij}^{1} are the result value after DE_{ij}, RE_{ij} is updated, $\max(DE_{ij})$, $\max(RE_{ij})$ are the maximum value under the sample, and DE_{ij}^{1}, RE_{ij}^{1} are the result value after DE_{ij}, RE_{ij} are updated.

Phases 2. Calculation of the weight of sample i for indicator j.

$$\rho_{ij}^{DE} = \frac{DE_{ij}^{1}}{\sum_{i=1}^{n} DE_{ij}^{1}} (i = 1, 2, \ldots, n; j = 1, 2, \ldots, m) \tag{3}$$

$$\rho_{ij}^{RE} = \frac{RE_{ij}^{1}}{\sum_{i=1}^{n} RE_{ij}^{1}} (i = 1, 2, \ldots, n; j = 1, 2, \ldots, m) \tag{4}$$

where ρ_{ij}^{DE}, ρ_{ij}^{RE} are the weight of the i sample in the j indicator; n and m are the number of rows and columns in the DE_{ij}^{1}, RE_{ij}^{1} sample, respectively.

Phases 3. Calculation of entropy value of the indicator. (j = 1, 2, …, m)

$$e_{j}^{DE} = -k \sum_{i=1}^{n} \rho_{ij}^{DE} * \ln(\rho_{ij}^{DE}) \tag{5}$$

$$e_{j}^{RE} = -k \sum_{i=1}^{n} \rho_{ij}^{RE} * \ln(\rho_{ij}^{RE}) \tag{6}$$

e_{j}^{DE}, e_{j}^{RE} are the entropy value of the indicator in column j, $k = \frac{1}{\ln(n)}$

Phases 4. Calculation of the coefficient of variation for the indicator

$$d_{j}^{DE} = 1 - e_{j}^{DE} \tag{7}$$

$$d_{j}^{RE} = 1 - e_{j}^{RE} \tag{8}$$

d_{j}^{DE}, d_{j}^{RE} are the coefficient of variation for indicator j.

Phases 5. Calculation of indicator weights

$$w_{j}^{DE} = \frac{d_{j}^{DE}}{\sum_{j=1}^{m} d_{j}^{DE}} (j = 1, 2, \ldots, m) \tag{9}$$

$$w_{j}^{RE} = \frac{d_{j}^{RE}}{\sum_{j=1}^{m} d_{j}^{RE}} (j = 1, 2, \ldots, m) \tag{10}$$

w_j^{DE}, w_j^{RE} are the weight of the j indicator and $\sum_{j=1}^{m} d_j^{\mathrm{DE}} \sum_{j=1}^{m} d_j^{\mathrm{RE}}$ are the sum of the coefficients of variation of all indicators in the system respectively.

3.2.2 Data Sources

This paper takes Jiaozuo City as the research object and collects the corresponding regional indicator data. Setting the time series of the study as 2011–2022, the data for each indicator in the digital economy indicator system mainly come from the database of the National Bureau of Statistics (NBS), the database of the National Research Network (NRN), and the EPS database, the relevant data under the digital manufacturing indicator come from the Henan Provincial Statistical Yearbook, and the digital inclusive finance index comes from the Digital Finance Research Center of Peking University. The data of the indicators in the real economy indicator system mainly come from the Henan Provincial Statistical Yearbook. For the unavailable individual data, this paper adopts the interpolation method to make up for the unavailable data.

3.2.3 Digital and Real Economy Development in Jiaozuo

According to the calculation of the entropy value method of the previous literature, the score of the level of digital economy and real economy in Jiaozuo City is calculated, and the score of economic development level of the real economy in Jiaozuo City is shown in Fig. 3.

The development level of the real economy in Jiaozuo City fluctuates from 0.33 in 2011 to 0.63 in 2022, and the development level of the digital economy increases from 0.06 in 2011 to 0.92 in 2022 which means that the development of real economy

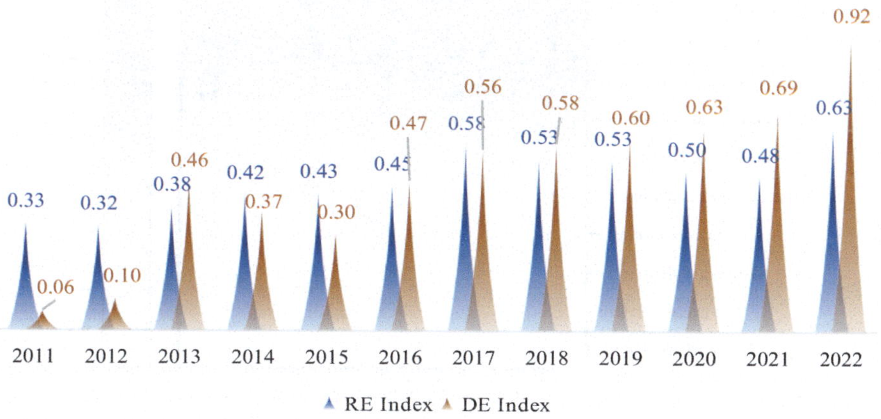

Fig. 3 Jiaozuo digital and real economies index

is relatively stable, while the development of digital economy is more rapid. In the *Work Program for the Development of Digital Economy in Henan Province in 2021*, the Henan government plans that in 2021, the growth of the digital economy in Henan Province should be above 13% and will further enhance the competitive advantage of the National Comprehensive Pilot Zone for Big Data.

Against this background, the digital economy in Jiaozuo City has developed relatively rapidly, and in November 2021, Jiaozuo City had basically achieved full coverage of 5G networks, and a number of digitalization projects such as the Jiaozuo Digital Economy Industrial Park had begun to be constructed or put into operation [40]. This is why the Jiaozuo Digital Economy Index grows from 0.69 to 0.92 in 2021–2022, a growth rate of 33%. Affected by the COVID19, the level of real economic development in Jiaozuo City has a tendency to wither from 2020–2021, and then rebound rapidly in 2022 with the end of the COVID19.

3.3 *Coupled Coordination Degree Model*

The Coupled Coordination Degree Model is extensively explored in various financial contexts. Studies analyze the coupling coordination degree between financial systems and enterprise development [28], regional economic development [11], green finance, digital economy, and ecological environment [30]. The coupling degree, also known as the correlation effect, is mainly used to analyze the correlation between two systems. The closer the value is to 1, the closer the two systems are [13]. The degree of coupling harmonization is classified into 10 levels are shown in Fig. 4 [43].

Based on the previous calculation of the digital and real economy, the study will use SPSSAU and the coupled coordination model to measure degree of synergy of

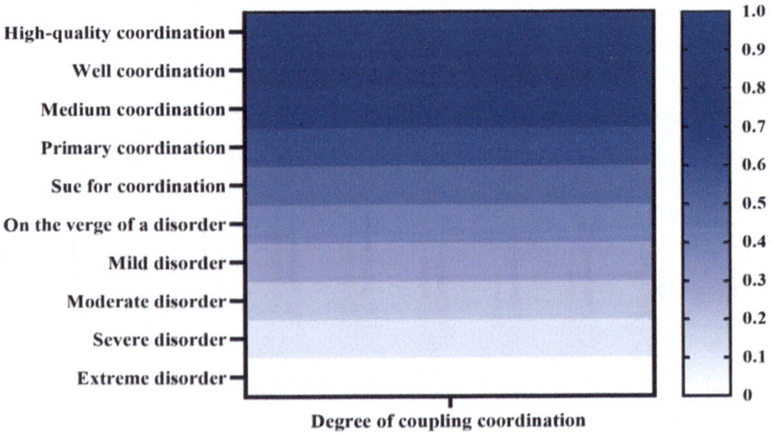

Fig. 4 Coupling evaluation criteria

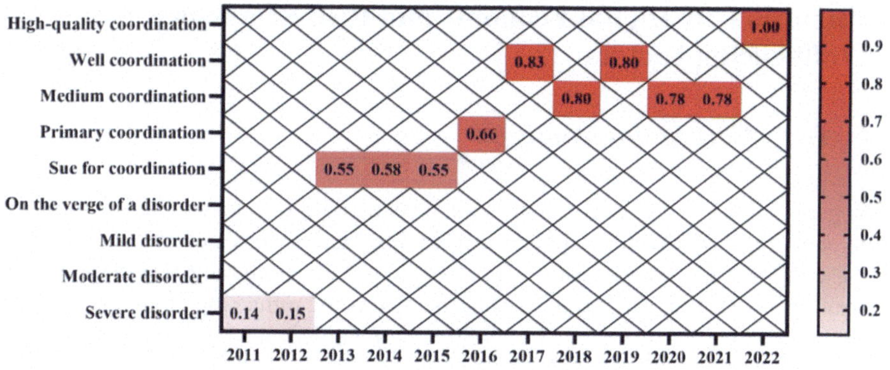

Fig. 5 Coupling degree between the real and the digital economies in Jiaozuo

the digital economy with the real economy in Jiaozuo. The results of the coupling degree between the real economy and the digital economy in Jiaozuo City are shown below in Fig. 5.

As shown in Figs. 4 and 5, the coupling degree between the digital economy and the real economy in Jiaozuo City has gradually increased from Severe disorder in 2011 to High-quality coordination in 2022. Meanwhile, the coupling coordination degree of digital economy and real economy in Jiaozuo City gradually increases from Severe disorder in 2011 to Well coordination in 2017, reflecting that the digital economy and the real economy have begun to integrate with each other. However, due to the backwardness of the digital economy infrastructure, the real economy has been slow to realize the effects of digitalization. The period from 2018 to 2022 is from medium coordination to high-quality coordination, and the development trend of the synergy of the two is more and more obvious.

This is due to the Jiaozuo government's emphasis on the development of synergy with digital and real economy. Jiaozuo government has increased the promotion of scientific and technological innovation, the implementation of the *ten major actions,* including the national high-tech zones to compete for advanced position, high-tech enterprises to multiply the lead, the full coverage of the R&D activities of industrial enterprises above the size of the industry chain to promote the industrial chain, the innovation chain, the supply chain, the factor chain, the depth of the coupling of the system chain [23].

4 Jiaozuo City Digital Economy and Real Economies Influencing Factors

4.1 Gray Correlation Analysis

This paper adopts the gray correlation analysis method to Jiaozuo City Digital and Real economy synergy development influencing factors, in which the parent order is the coupling and coordination degree of digital economy and real economy in Jiaozuo from 2011 to 2022, and the subsequence is the influencing factors. The real economy is the pillar of society. Its vitality and resilience are the important support for the sound operation of the social and economic system.

The level of economic efficiency and economic development indeed mirrors the advancement of the real economy and holds a crucial position in the progression of digital and real economies synergy [29]. Disposable income per capita is indeed a crucial indicator of economic development, reflecting the average money available to individuals after tax deductions [4]. Moreover, the number of industrial enterprises above scale is also an important indicator of economic efficiency, reflecting the innovation performance and overall innovation level of the industry [31]. The degree of openness in a region plays a vital role in its capacity to integrate external resources, particularly for the adoption of digital technology and market expansion. Foreign direct investment (FDI) is often utilized as an indicator of this openness, reflecting the region's attractiveness to external investors and its ability to engage with global markets [34]. Research and development (R&D) investment plays a crucial role in driving innovation in the digital economy [3, 39]. Studies have shown that R&D expenditure positively impacts innovative investment and green innovation capability, especially in regions with a strong digital economy presence [3]. The percentage of people in the workforce indeed reflects the structure of the industry, as evidenced by various studies. For instance, the shift from goods-producing to service-producing sectors has been significant, with the latter now accounting for a larger share of employment, thereby stabilizing total employment during economic downturns [3].

In summary, the digital and real economy synergy of development factors affecting the variables of the level of economic development, economic efficiency, openness, industrial structure and R & D investment; variable indicators are determined as the disposable income per capita, number of industrial units above designated size, Foreign direct investment, Percentage of people in the workforce and Internal expenditure on R&D funds. The indicators of factors influencing the development of digital and real economies synergy are shown as Fig. 6.

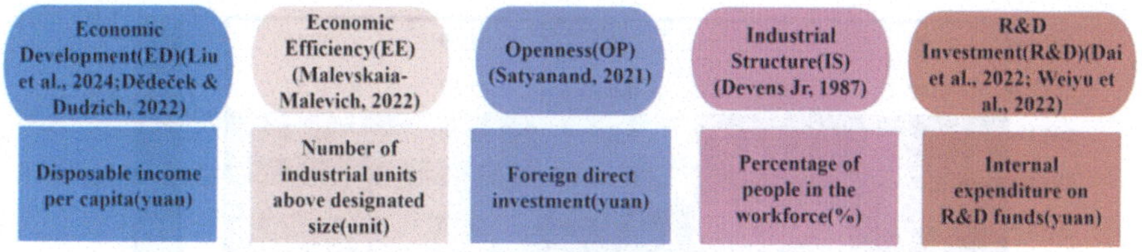

Fig. 6 Indicators of factors influencing the development of digital and real economies synergy selection

4.2 *Results of Empirical Analysis*

The correlation coefficients of Jiaozuo City's digital and real synergy development level and influencing factors, as well as the correlation degree of each influencing factor are calculated according to the gray correlation analysis method through SPSSAU software as shown below the Figs. 7 and 8. By comparing the size of the correlation degree, the closeness of the influence of each influencing factor on the development level of digital and real synergy in Jiaozuo City can be derived. Generally speaking, The correlation coefficient has a value between 0 and 1, the more the value is, the more the level of Jiaozuo's digital and real synergy development will be affected. When the correlation coefficient is between 0 and 0.35, it indicates the weakest correlation; when the correlation coefficient is between 0.35 and 0.65, it is moderately correlated; when the correlation coefficient is between 0.65 and 0.85, the correlation is strong; when the correlation coefficient is between 0.85 and 1 time, the correlation is extremely strong, and the law of interaction between the research indicators is basically the same [2].

From the empirical results, it is clear that the development of the real economy in Jiaozuo City is lagging behind the development of the digital economy, and the slow down of the real economy under the development of digitization will affect the further enhancement of digitization, which may lead to the development of digital-real synergy in Jiaozuo City to produce a vicious circle. To narrow the gap between the real economy and the digital economy, it is essential to focus on integrated development, leveraging the immense potential of digital technology while strengthening the foundations of the traditional economy [27]. The real economy of Jiaozuo City, especially the industrial sector, is realizing high-quality development through scientific and technological innovation and industrial upgrading. For example, DuoFuDuo New Materials Co., Ltd. has achieved improved productivity and reduced energy consumption through digital transformation. Meanwhile, it is accelerating the implementation of green manufacturing upgrading actions, utilizing digital technology to promote energy saving and carbon reduction, developing green factories and green parks, and constructing a green manufacturing system for the whole industrial chain. In terms of digital infrastructure, Jiaozuo has continued to improve the construction of 5G infrastructure to achieve continuous coverage in areas above townships, and

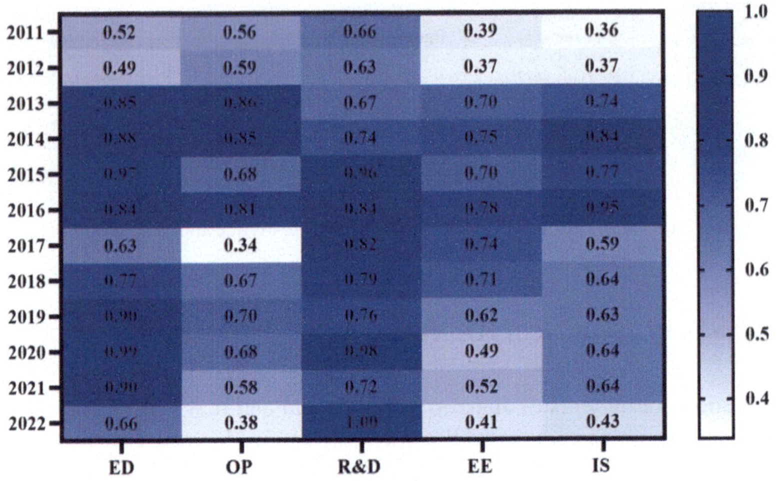

Fig. 7 Correlation coefficient of factors influencing digital and real economies synergy from 2011 to 2022

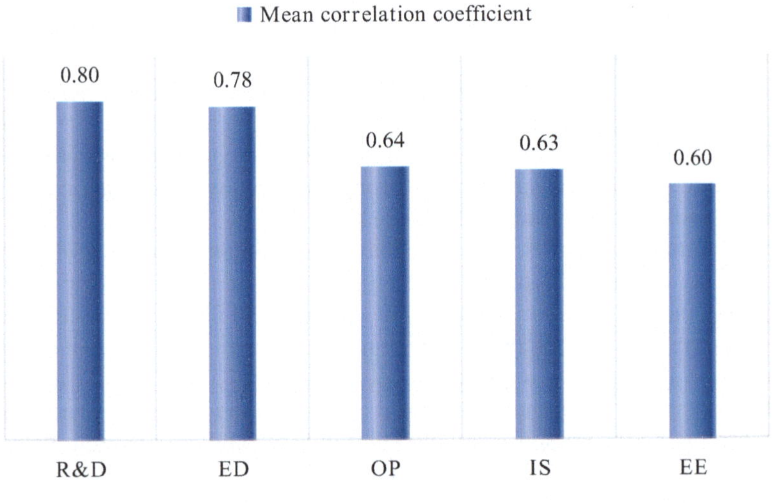

Fig. 8 Mean correlation coefficient of factors influencing digital and real economies synergy

has promoted the construction of a big data service platform to support the synergy of the digital and real economies.

As shown in Figs. 7 and 8, the correlation of R&D investment and Economic Development is 0.80 and 0.78 which indicating their strong correlation with the development level of digital and real economies synergy, The correlations of the remaining indicators such as Openness, Economic Efficiency and Industrial Structure are 0.64, 0.60 and 0.63 respectively, indicating that these indicators are moderately correlated with Jiaozuo City's digital and real economy synergy.

The strength of R&D investment has the most obvious impact on the development of digital and real synergy in Jiaozuo. Research and development (R&D) investment plays a pivotal role in the synergy of the digital and real economies, significantly influencing their development. The digital economy, characterized by technologies such as big data, cloud computing, and artificial intelligence, drives high-quality economic growth and optimizes resource allocation [7]. R&D innovation is crucial in this context, as it helps overcome technological barriers and enhances core competitiveness, thereby fostering the digital transformation of various sectors like agriculture, manufacturing, and services [25]. By increasing investment in R&D, Jiaozuo City has promoted the digitalization and intelligent transformation of traditional industries, improving production efficiency and product quality. This not only promotes industrial upgrading, but also promotes technological innovation, especially in the fields of new materials, new energy, biomedicine and other strategic emerging industries, meanwhile, Jiaozuo Government encourages enterprises to increase R&D investment, and pushes industrial enterprises above the scale to dock with at least one university or scientific research institute, set up an R&D team led by high-level talents, and build an R&D platform, so as to stimulate the innovation power of enterprises. R&D investment in Jiaozuo City, the synergy of digital and real economy has played a key role in promoting industrial upgrading and high-quality economic development [24].

The level of economic development also has a strong correlation to the development of digital and real synergy in Jiaozuo City. With the increase of residents' income, Jiaozuo can further optimize the economic structure, increase the investment in the core industries of digital economy, and improve the digitization and intelligence level of the industry. The growth of per capita disposable income is directly related to the improvement of the quality of life of the residents, which helps to increase the acceptance and participation of Jiaozuo residents in the digital life services, and the increase of income level can stimulate the enthusiasm of the residents for employment and entrepreneurship, especially in the field of digital economy, providing talent and innovation support for digital real synergy.

The correlations of indicators such as Openness, Economic Efficiency and Industrial Structure are moderately correlated with Jiaozuo City's digital and real economy synergy. The degree of openness influences the digitization process of Jiaozuo City, and the entry of foreign enterprises to a certain extent promotes the development of e-commerce, cloud computing and artificial intelligence, among other fields. The influence of industrial structure on the development of Jiaozuo City's digital and real synergy is mainly manifested. Labor force, as an indispensable driving force in the process of synergy development, continues to have a positive impact on the development of digital and real economy synergy.

5 Conclusion

In this paper, on the basis of exploring the digital and real economies synergy of Jiaozuo City, the entropy value method and the coupling coordination degree model are used to measure the development level of digital and real economies synergy in Jiaozuo City, and the gray correlation analysis method is used to explore the influencing factors of the digital and real economies synergy development in Jiaozuo City, which will provide a reference to the development of digital and real economies synergy in other prefectural-level cities. The study finds that the digital economy and real economy of Jiaozuo City have steadily developed from a Severe disorder in 2011 to a high-quality coordination in 2022. However, the development of the real economy has shown a trend of lagging behind the development of the digital economy, the level of economic development and R&D investment have the most significant impact on the development of digital and real economies synergy, whereas other influences such as Openness, Economic Efficiency, Industrial Structure and Economic Development have different effects in Jiaozuo City.

In summary, it can be seen that the level of economic development and R&D investment have the highest impact on the synergy of digital and real economy in Jiaozuo City. Therefore, government departments should increase their efforts to stimulate enterprises' R&D investment. The government should formulate relevant policies to support R&D investment, such as fiscal and tax incentives, financial support and incentives for scientific and technological innovation, in order to provide enterprises with a better environment and policy support for R&D investment. Meanwhile, Jiaozuo has to continue to promote the cooperation between enterprises and universities and research institutes, enhance the R&D capability of enterprises by building R&D institutions and carrying out joint R&D projects, implement the *Jiaozuo Talent Program* to introduce and cultivate high-level talents to provide talent support for enterprises, and evaluate and notify the R&D activities of enterprises in Jiaozuo City, so as to motivate enterprises to Increase investment in R&D.

References

1. Central People's Government.: Statistical classification of the digital economy and its core industries (2021). https://www.gov.cn/gongbao/content/2021/content_5625996.htm
2. Chen, X.: (2023). Gray correlation analysis. In: Chen, X. (ed.) Application of Gray System Theory in Fishery Science, pp. 35–75. Springer Nature Singapore. https://doi.org/10.1007/978-981-99-0635-2_3
3. Dai, D., Fan, Y., Wang, G., Xie, J.: Digital economy, R&D investment, and regional green innovation—analysis based on provincial panel data in China. Sustainability **14**(11), 6508 (2022). https://doi.org/10.3390/su14116508
4. Dědeček, R., Dudzich, V.: Exploring the limitations of GDP per capita as an indicator of economic development: a cross-country perspective. Rev. Econ. Perspec. **22**(3), 193–217 (2022). https://doi.org/10.2478/revecp-2022-0009

5. Ding, C., Liu, C., Zheng, C., Li, F.: Digital economy, technological innovation and high-quality economic development: based on spatial effect and mediation effect. Sustainability **14**(1), 216 (2022). https://doi.org/10.3390/su14010216
6. Du, Y.-C., Jiang, P., Mao, J.-S., Xu, L.-Y.: Eco-efficiency of Jiaozuo industry sectors. Huanjing kexue **32**(5), 1529–1536 (2011). http://europepmc.org/abstract/MED/21780616
7. Fu, L.: Research on the mechanism of the influence of digital economy on the high-quality development of industry. Scien. Soc. Res. **6**(6), 270–276 (2024). https://doi.org/10.26689/ssr.v6i6.7438
8. Fu, Y., Yu, H., Da, J., Kan, X.: Discussion on the influence of digital economy development on real economy. Front. Bus. Econ. Manage. **6**(3), 176–179 (2022). https://doi.org/10.54097/fbem.v6i3.3458
9. Gao, Z., Lu, Z.: An empirical study on the integrated development of equipment manufacturing and high-tech service industries under the theory of system coupling. J. Syst. Sci. **27**(2), 63–68 (2019)
10. Grilli, L., Santoro, D.: Forecasting financial time series with Boltzmann entropy through neural networks. CMS **19**(4), 665–681 (2022). https://doi.org/10.1007/s10287-022-00430-2
11. Gu, R.: Innovative paths of modern fiscal and financial systems to assist the coordinated development of regional economy. Appl. Math. Nonlinear Sci. **9**(1), 1–14 (2024). https://doi.org/10.2478/amns.2023.2.01265
12. Guo, H., Quan, Q.: Integration of the digital economy and the real economy: Measurement, evaluation, and implementation paths. Econ. Rev. **11**, 72–82 (2022). https://doi.org/10.16528/j.cnki.22-1054/f.202211072
13. Han, Z., Wu, H., Zhao, J.: Coupling coordination measurement and evaluation of digital economy and high-quality development. J. Statis. Inf. **6**(37), 22–34 (2022). https://doi.org/10.1371/journal.pone.0259132
14. Henan Provincial Bureau Statistics.: Statistical Yearbook (2023). https://tjj.henan.gov.cn/tjfw/tjcbw/tjnj/
15. Hong, Z., Xiao, K.: Digital economy structuring for sustainable development: the role of blockchain and artificial intelligence in improving supply chain and reducing negative environmental impacts. Sci. Rep. **14**(1), 3912 (2024). https://doi.org/10.1038/s41598-024-53760-3
16. Institute of Digital Finance Peking University.: Peking University Digital Inclusive Finance Index (2021). https://idf.pku.edu.cn/zsbz/index.htm
17. Jiao, S., Sun, Q.: Measurement of digital economy development in China and its influencing factors. World Surv. Res. (7), 13–23 (2021). https://doi.org/10.13778/j.cnki.11-3705/c.2021.07.002
18. Jiaozuo Human Resources and Social Security Bureau.: Jiaozuo talent attraction action public introduction of high-level and urgently needed shortage of talent announcement (2023). http://rsj.jiaozuo.gov.cn/sitesources/jzsrsj/page_pc/zwxx/bmdt/article891c488fa825444585fbfc0b78b5afe5.html
19. Jiaozuo Municipal Committee.: The 14th five-year plan for national economic and social development of Jiaozuo city (2021). https://www.henan.gov.cn/2021/01-26/2085884.html
20. Jiaozuo Municipal Government.: New intelligent city construction implementation program in Jiaozuo (2021). http://www.jiaozuo.gov.cn/sitesources/jiaozuo/page_pc/zwgk/szfxxgk/zfwk/szfbwj/articlefd868c91a49143b787b1878070ebcd61.html
21. Jiaozuo Municipal Government.: Jiaozuo city strengthening digital government construction implementation program (2024). http://www.jiaozuo.gov.cn/sitesources/jiaozuo/page_pc/zwgk/szfxxgk/zfwk/szfwj/articled58919e162d14b02ae7c90654355b654.html
22. Jiaozuo Municipal Government.: Work report 2024 (2024). http://www.jiaozuo.gov.cn/sitesources/jiaozuo/page_pc/zwgk/szfxxgk/zfgzbg/article9623d796166648b8ae8fa860ee4db77d.html
23. Jiaozuo Municipal People's Government.: The first power again Jiaozuo City to promote scientific and technological innovation work (2022). https://www.henan.gov.cn/2022/03-10/2411833.html

24. Jiaozuo Municipal People's Government.: Jiaozuo City promotes science and technology innovation (2022). https://www.henan.gov.cn/2022/03-10/2411833.html
25. Jin, M.: Research on the impact and opportunities of the digital economy promoting the development of real economy. Proc. Bus. Econ. Stud. **7**(2), 126–131 (2024). https://doi.org/10.26689/pbes.v7i2.6777
26. Kemaev, K.V., Kemaeva, M.V., Grinevich, Y.A.: The development of the real sector as the basic factor of the sustainability of regional economic systems. Scien. Res. Dev. Econ. (11), 16–21 (2023). https://doi.org/10.12737/2587-9111-2023-11-1-16-21
27. Li, X.: The path of integrated development between digital economy and real economy. Acad. J. Bus. Manage. **5**(10), 160–164 (2023). https://doi.org/10.25236/AJBM.2023.051025
28. Lin, X., Chen, G.: Path analysis of financial assistance to business development by applying coupled coordination degree model. Appl. Math. Nonlinear Sci. **9**(1), 1–13 (2024). https://doi.org/10.2478/amns.2023.2.00920
29. Liu, B., Huang, Y., Chen, M., Lan, Z.: Towards sustainability: how does the digital-real integration affect regional green development efficiency? Econ. Anal. Policy **83**, 42–59 (2024). https://doi.org/10.1016/j.eap.2024.06.008
30. Liu, Z., Zhang, X., Wang, J., Shen, L., Tang, E.: Evaluation of coupling coordination development between digital economy and green finance: evidence from 30 provinces in China. PLoS ONE **18**(10), e0291936 (2023). https://doi.org/10.1371/journal.pone.0291936
31. Malevskaia-Malevich, E.D.: Development of performance indicators for sustainable enterprises. Вестник университета **1**(11), 138–144 (2022). https://doi.org/10.26425/1816-4277-2022-11-138-144
32. National Internet Information Office.: Digital China development report (2022). Beijing People's Publishing House (2023). https://www.cac.gov.cn/2023-05/22/c_1686402318492248.htm
33. Orozonova, A., Zhang, S., Zhumadilov, E., Sun, X., Liu, X.: Modelling the impact of the digital economy on the development of the logistics industry. study case of Henan province. Theor. Pract. Res. Econ. Fields **15**(2), 232–244 (2024). https://doi.org/10.14505/tpref.v15.2(30).07
34. Satyanand, P.N.: Foreign direct investment and the digital economy (2021). https://hdl.handle.net/20.500.12870/3730
35. Tian, K., Xia, X., Lu, C., Peng, B.: Research on the digital economy boosting the high-quality development of manufacturing in underdeveloped areas—taking Kaifeng city as an example. Open J. Bus. Manage. **10**(6), 3048–3066 (2022). https://doi.org/10.4236/ojbm.2022.106151
36. Wang, C., Liu, Y., Wan, Y., Hu, S., Xia, H.: How does the digital economy impact the green upgrading of manufacturing? Perspectives on technological innovation and resource allocation. Appl. Econ., 1–16 (2024). https://doi.org/10.1080/00036846.2024.2364112
37. Wang, W., Mei, T.: Research on the effect of digital economy development on the carbon emission intensity of agriculture. Sustainability **16**(4), 1457 (2024). https://doi.org/10.3390/su16041457
38. Wei, X., Zhang, J., Liu, Y.: Digital economy and farmers' income inequality—a quasi-natural experiment based on the "broadband China" strategy. Appl. Econ. Lett., 1–6. https://doi.org/10.1080/13504851.2024.2332553
39. Weiyu, Z., Othman, A., Guli, T.: Role of technological acquisition and R&D expenditure in innovative investment. Front. Psychol. **13,** 855295 (2022). https://doi.org/10.3389/fpsyg.2022.855295
40. Xu, W.: "Digital Jiaozuo" future is promising. Jiaozuo Daily (2021)
41. Yan, J., Lu, W., Xu, X., Lian, J.: Empirical study of the environmental Kuznets curve in China based on provincial panel data. Sustainability **15**(6), 5225 (2023). https://doi.org/10.3390/su15065225
42. Zhang, D.: Exploring the trends of the relationship between digital industrialization and industrial digitization on the development of digital economy—take China and Japan as examples. Adv. Econ. Manage. Polit. Sci. **82**, 287–294 (2024). https://doi.org/10.54254/2754-1169/82/20230992

43. Zhou, D., Feng, C.: The coupling and interaction relationship between science and technology finance and high-quality economic development: an empirical analysis based on coupling degree and PVAR modeling. Techno-economic **39**(5), 107–115 (2020). CNKI:SUN:JSJI.0.2020-05-014
44. Zhou, H., Li, X., Li, X.: Can the digital economy improve the level of high-quality financial development? Evidence from China. Sustainability **15**(9), 7451 (2023). https://doi.org/10.3390/su15097451

Strategic Marketing and Customer Analytics

Jonnalagadda Anil Kumar and S. Ramesh Babu

Abstract Background: Strategic Marketing and Customer Analytics aims to explore the integration of marketing strategy with data analytics to enhance customer-centric approaches. It emphasizes the importance of aligning marketing initiatives with organizational goals and delves into the fundamentals of strategic marketing. Objectives: The chapter proposes to investigate how data analytics can be utilized to drive customer-centric strategies in marketing. Specifically, it aims to explore behaviour analysis, segmentation, and predictive modelling within the realm of customer analytics. Methods: The chapter examines real-world cases that demonstrate the effective utilization of analytics in personalizing campaigns, optimizing customer acquisition, and managing customer relationships. The methods employed include analysing behaviour patterns, segmenting target audiences, and developing predictive models based on data insights. Results: The results revealed the potential of leveraging data analytics to enhance marketing strategies. Real-world cases illustrate how analytics can be used to personalize campaigns, optimize customer acquisition processes, and foster stronger customer relationships. These insights empower marketers to make informed decisions and stay relevant in a dynamic market environment. Conclusion: The integration of marketing strategy with data analytics offers significant opportunities for organizations to drive customer-centric approaches. By aligning marketing initiatives with organizational goals and leveraging analytics effectively, marketers can enhance their competitiveness and relevance in the marketplace. This chapter underscores the importance of data-driven decision-making in ensuring marketing success and organizational sustainability.

Jonnalagadda Anil Kumar (✉) · S. Ramesh Babu
Department of Commerce and Management, Koneru Lakshmaiah Education Foundation, Green Fields, Vaddeswaram, Guntur, Andhra Pradesh 522502, India
e-mail: janilkumar@yahoo.com

S. Ramesh Babu
e-mail: srb.rameshbabu@kluniversity.in

Jonnalagadda Anil Kumar
Centurion School of Smart Agriculture, Centurion University of Technology and Management, Vizianagaram, Andhra Pradesh 535003, India

S. Kautish et al. (eds.), *Strategy Analytics for Business Resilience Theories and Practices*, Information Systems Engineering and Management 33,
https://doi.org/10.1007/978-3-031-82369-5_5

Keywords Strategic marketing · Customer analytics · Data analytics · Customer-centric approach · Predictive modelling

1 Introduction

1.1 Overview of Strategic Marketing

Effective marketing strategies are essential for businesses to achieve their goals by targeting the right audience, communicating value propositions, and standing out in a competitive environment. Understanding customer needs drives product development, pricing, distribution, and promotions. Modern strategies must adapt to market trends, blending traditional and digital methods for maximum reach and engagement. Data analytics play a crucial role in customization, improving relevance, impact, satisfaction, and loyalty. Integrating customer insights allows organizations to stay ahead of market dynamics and maintain a competitive edge. Strategic marketing with data analytics enables resource optimization, success measurement, and informed decision-making. By prioritizing customer-focused approaches supported by analytics, businesses can ensure long-term growth and success in a digital landscape [1, 2].

1.2 Importance of Customer Analytics

Customer analytics is crucial in strategic marketing, changing how businesses make decisions and navigate the market. Data-driven decision-making is essential for success, with tools enabling companies to gather and interpret data for optimized strategies.

Analysing customer preferences through analytics helps businesses personalize recommendations, leading to higher engagement and ROI. Real-time monitoring of campaign performance allows for continuous refinement and improvement for maximum impact.

Predictive analytics using advanced algorithms helps companies anticipate future trends and consumer behaviour, empowering them to make confident decisions based on customer insights. Ethical data practices and privacy are important considerations to build trust with customers and comply with regulations.

In summary, integrating customer analytics into strategic marketing is key for sustainable success in today's data-driven world. By embracing evidence-based decision-making processes, businesses can enhance targeting strategies, optimize campaigns efficiently, predict future trends accurately, and foster strong relationships with customers based on trust and transparency. See references: Basu et al. [3], Noha [4], Rosário and Dias [5].

1.3 *Objectives of the Report*

The aim of this document is to explore the convergence of strategic marketing and customer analytics, delving into how data-driven insights can shape and improve marketing strategies. By synthesizing current knowledge and identifying research gaps, this document seeks to offer a comprehensive overview of the changing landscape of marketing analytics. Through a thorough analysis of publication trends, top contributors, and research themes in marketing analytics, this document aims to provide valuable insights for both academics and industry professionals. Moreover, this document will investigate the role of marketing analytics in guiding strategic long-term decision-making across a variety of sectors beyond just retail and services.

By addressing crucial research inquiries such as the influence of marketing analytics on operational decisions like campaign management and market segmentation, this document strives to measure the efficacy of data-driven approaches in driving business results. Additionally, this report will emphasize the shift of marketing analytics from transactional functions to strategic orientations, underscoring the necessity for innovative practices that can confront future obstacles. By merging aspects of customer behaviour with data analytics, this report aspires to offer a more comprehensive understanding of customer decision-making processes. Ultimately, the objectives of this report are to pave the way for inventive marketing strategies that leverage analytics for tailored campaigns, optimized acquisition tactics, and improved customer relationships based on data-backed decisions [3].

2 Fundamentals of Strategic Marketing

2.1 *Defining Strategic Marketing*

Marketing strategy involves aligning company objectives with marketing tactics to achieve business success through analysing market trends, consumer behaviour, and competition. It focuses on seizing market opportunities by leveraging strengths and addressing weaknesses for sustainable growth and competitive advantage. With the use of technologies like artificial intelligence (AI), big data, and machine learning, data analytics has transformed marketing strategy by providing a greater understanding of the preferences, behaviours, and choices of customers. This enables businesses to target specific audiences, predict demand accurately, and enhance customer interaction.

Customer analytics is crucial in segmenting customers based on preferences to tailor marketing strategies effectively, boosting satisfaction, fostering loyalty, and increasing profitability. Developing focused marketing materials that resonate with consumers and spur business expansion necessitates matching organisational objectives with data-driven insights. By understanding customer desires and preferences through analytics, businesses can develop innovative strategies to stand out in the competitive market [3].

2.2 Relationship with Organizational Goals

The correlation between strategic marketing and corporate objectives is crucial for the prosperity of any business. Strategic marketing entails harmonizing marketing strategies with the overarching goals and vision of the organization to propel growth and profitability. By outlining clear objectives and tactics, businesses can position themselves effectively in the marketplace, set their offerings apart, and establish lasting competitive advantages.

The organizational goals act as a guiding framework for strategic marketing endeavours, influencing decisions on development of products, strategies for pricing, channels of distribution, and promotional campaigns. By leveraging data-driven decision-making, businesses can optimize their marketing initiatives, personalize customer engagements, and enhance customer acquisition and retention.

Ultimately, the alignment between strategic marketing and organizational objectives ensures that marketing efforts are not isolated but are closely intertwined with the broader goals of the business. Companies can increase revenue growth, strengthen their brand recognition in the marketplace, and optimise the efficacy of their marketing initiatives thanks to this alignment [1].

2.3 Key Concepts in Strategic Marketing

In essence, strategic marketing emphasises the value of marketing analytics in making well-informed decisions. To identify market trends and inform marketing strategy, marketing analytics entails gathering, evaluating, and presenting data using a variety of technologies. It plays a vital role in aligning marketing efforts with organizational goals, improving customer engagement, and optimizing acquisition strategies. By integrating data analytics into marketing approaches, companies can personalize campaigns, segment customers efficiently, and forecast customer behaviour for enhanced engagement.

Furthermore, strategic marketing entails utilizing analytics for tailored campaigns and nurturing customer connections through data-informed choices. It concentrates on cultivating lasting customer relationships by utilizing data to enhance the overall customer journey. This approach improves client satisfaction and loyalty while also strengthening acquisition methods. An integral part of strategic marketing is assessing customer behaviour through behavioural analysis tools and methodologies.

Essentially, essential concepts in strategic marketing revolve around utilizing data analytics to steer decision-making processes, enhance customer interactions, and achieve organizational goals effectively. By adopting these concepts, businesses can maintain competitiveness in today's ever-changing marketplace and adapt to the evolving landscape of digital marketing tactics. See references: Basu et al. [3], GGI Insights [1].

3 Integration of Data Analytics in Marketing Strategy

3.1 Understanding Data Analytics

Utilizing data analytics is a game-changer in revolutionizing strategic marketing, as it offers businesses valuable insights obtained from analysing extensive data sets. By incorporating cutting-edge analytical tools like artificial intelligence, big data, and machine learning, organizations can unlock new potential in their operations. Through the thorough examination of large data volumes, businesses can make well-informed decisions that contribute to long-term enhancements in performance.

For instance, by examining historical sales data and market patterns, the application of big data analytics increases the accuracy of demand forecasts in the fast-moving consumer products sector. This forward-thinking approach optimizes inventory management, resulting in greater consumer satisfaction and cost efficiencies. Similarly, AI-driven predictive analytics are revolutionising customer interactions in retail by using customer behaviour and demographic data to create personalised marketing materials and product recommendations.

Moreover, the application of machine learning and big data in financial services aids in identifying market trends and patterns, enabling firms to accurately predict market movements. In the hospitality sector, predictive analytics refine pricing strategies and capacity management for increased profitability. Across diverse industries, the integration of analytics transforms business operations while opening new avenues for expansion and profitability.

In summary, utilizing the power of data analytics in strategic marketing enables businesses to make informed decisions that foster performance enhancements and sustainable growth. See references: Basu et al. [3], The Importance of Data Analytics in Digital Marketing [6].

3.2 Benefits of Integrating Data Analytics in Marketing Strategy

Incorporating data analytics into marketing strategies presents a plethora of advantages that can significantly boost the efficacy of campaigns. By utilizing advanced analytical tools, marketers can gain valuable insights into the performance of digital marketing initiatives at all stages of the customer journey. Data analytics helps businesses eliminate guesswork from their marketing plans and optimize resource allocation, enabling a more personalized approach to customer interactions that enhances satisfaction and loyalty.

One major benefit of integrating data analytics is the ability to delve deeper into understanding customer behaviour. Businesses can tailor their branding, product strategy, and marketing initiatives to their target market by examining large amounts of customer data. This results in more efficient digital marketing techniques and personalized customer engagements, ultimately driving increased efficiency and profitability for the business.

Furthermore, consolidating data from diverse sources provides a holistic view of market trends and customer behaviours. This facilitates the development of targeted and personalized marketing strategies tailored to specific customer segments, thereby boosting relevance and engagement. By breaking down information silos and integrating data from various sources, businesses can create a unified perspective of the customer journey, leading to more informed decision-making in marketing endeavours.

To sum up, integrating data analytics into marketing strategies brings forth numerous benefits such as enhanced comprehension of customer behaviour, targeted marketing tactics, personalized customer interactions, and informed decision-making. By improving the capabilities of data processing, organizations can streamline their digital marketing campaigns and elevate overall business performance [7].

3.3 Challenges and Considerations

When it comes to strategic marketing and customer analytics, the incorporation of data analytics presents various obstacles and factors for businesses to consider. A major obstacle is how to incorporate big data analytics into decision-making procedures in an efficient manner. The inability to handle and analyse massive amounts of data is caused by a lack of internal experience, which calls for statistical expertise, programming abilities, and a thorough understanding of business processes. Tasks like data cleaning, integration, and analysis from different sources require specific technical abilities. Moreover, the absence of a comprehensive approach to analytics impedes smooth integration across the decision-making process. Many companies view analytics as a separate function rather than an essential component of decision-making, making it difficult to establish a cohesive strategy.

To tackle these challenges, businesses should prioritize enhancing internal expertise and fostering a culture that values data-driven insights as critical elements of strategic decisions. By investing in training programs and tools that support the integration of analytics into decision-making processes, organizations can improve their marketing effectiveness and overall business results.

Businesses may fully realise the potential of consumer analytics to drive growth and success by taking these factors into account and actively working to overcome obstacles linked to integrating data analytics into marketing strategy [3, 8].

4 Customer Behaviour Analysis

4.1 Importance of Understanding Customer Behaviour

A crucial component of customer analytics and strategic marketing is capturing consumer behaviour. Through investigating the needs, requirements, purchasing path, and pain points of their clients, businesses can obtain significant insights that inform their decision-making procedures. Customer analytics enables the gathering and examination of data concerning how customers engage with a brand, website, and products. This data is crucial in forecasting and reacting to shifts in customer behaviour, improving overall customer satisfaction and revenue, as well as executing targeted sales and marketing strategies.

Through the analysis of customer behaviour, businesses can pinpoint trends in customer data that offer insight into their preferences, motivations, and buying habits. Using several tools and techniques for behaviour research, companies can customise their marketing strategies to cater to the unique requirements of different client segments.

Segmentation is essential in comprehending customer behaviour as it assists in grouping customers based on shared traits. This makes it possible to send promotions and marketing messages that are specifically tailored to each segment's tastes and behaviour. Predictive modelling further boosts customer engagement by predicting future purchasing behaviours based on past data.

In essence, grasping customer behaviour through analytics empowers businesses to optimize acquisition tactics, cultivate lasting relationships with customers through data-informed decisions, and enhance the overall customer experience. By leveraging insights gleaned from the analysis of customer behaviour, companies can devise personalized campaigns that address individual needs while fostering growth and loyalty [3, 9, 10].

4.2 Tools and Techniques for Behaviour Analysis

Understanding customer behaviour is crucial in marketing, offering insights into preferences and decision-making. Combining traditional research with data mining helps track engagement stages for effective relationship management. Sentiment analysis uncovers emotional triggers impacting behaviour. Customer behaviour analytics platforms provide insights on interactions, allowing tailored marketing messages.

Data-driven marketing enables personalized experiences through customer data and analytics, improving conversion rates and satisfaction levels. These tools and techniques are essential for strategic marketing by creating customized campaigns, refining acquisition strategies, fostering relationships, and enhancing the overall customer experience [3, 5, 10].

5 Customer Segmentation

5.1 Purpose and Benefits of Segmentation

A key marketing tactic is customer segmentation, which entails putting customers in groups according to traits they have in common, such as preferences, behaviours, and demographics. Segmentation serves the primary purpose of personalising marketing techniques for consumer groups, enabling companies to develop focused advertising campaigns that appeal to their target market. Businesses can better understand their diverse client base and send more engaging and conversion-boosting messaging by segmenting their consumer base.

Enhancing brand recognition, client retention, and generating new sales are some of the main benefits of consumer segmentation. Businesses may better understand the requirements and interests of their customers by grouping them according to shared characteristics. This facilitates communication and fosters stronger relationships. This focused approach also enables companies to provide more targeted advertising and adjust product features and pricing based on segment-specific insights.

Furthermore, customer segmentation helps companies identify valuable customers, optimize acquisition strategies, and implement personalized retention tactics. Through analyzing customer data using segmentation techniques, businesses can identify opportunities for growth, enhance products or services, and launch tailored marketing campaigns to re-engage undecided or lost customers. For example, by utilizing the RFM model (Recency, Frequency, Monetary Value), companies can prioritize marketing efforts across different stages of the customer journey—from attracting leads to retaining loyal customers.

In conclusion, customer segmentation is a useful approach that helps companies better understand their clients and cater to their demands. By grouping customers into meaningful segments and adjusting marketing strategies accordingly, companies can build stronger relationships with their audience, boost brand loyalty, and drive sustainable business growth. See reference Avasari [11].

5.2 Types of Segmentation Strategies

Partitioning consumers into discrete groups according to common attributes such as demographics and behaviours is an essential marketing strategy. By using this technique, companies can get more insight into their clientele and deliver offers and communications that are specifically designed to affect consumers' decisions to buy. Through consumer segmentation based on variables like industry, firm size, region, and purchasing patterns, marketers may fine-tune their messaging and targeting tactics for a more effective outreach. For example, categorizing customers based on their position in the customer lifecycle—whether they are leads, new customers, loyal patrons, at-risk clients, or churned customers—enables businesses to tailor their marketing efforts accordingly.

Utilizing the RFM model (Recency, Frequency, Monetary Value) to assess customers and identify high-value clientele allows companies to focus on segments with greater potential for business growth. Additionally, customer segmentation contributes to boosting brand awareness, customer retention, and generating new sales through improved ad targeting and product enhancements. By setting clear goals and objectives to guide marketing efforts effectively, businesses can ensure a more targeted approach in reaching their audience.

Customer segmentation methods may range from demographic and geographic segmentation to psychographic or behavioural segmentation based on technological insights or customer status reflecting their position in the customer lifecycle. However, it is crucial to avoid creating overly narrow segments that do not evolve over time.

In the end, knowing client behaviour patterns and preferences requires efficient customer segmentation. It gives companies the ability to modify their marketing plans in response to the requirements of various client segments, increasing relevance and engagement and fostering enduring bonds through individualised interactions [7, 11, 12].

6 Predictive Modelling for Customer Engagement

6.1 Overview of Predictive Modelling

Prognosticating modelling stands as a game-changer in the realm of marketing, revolutionizing strategies by offering foresight into upcoming trends and behaviours. Through the utilization of data mining, predictive modelling, and machine learning, marketers can forecast marketing trends, customer behaviour, and campaign outcomes. This forward-thinking methodology empowers businesses to predict customer needs, customize marketing strategies efficiently, and cultivate brand loyalty. Predictive analytics in marketing empowers marketers to craft personalized campaigns that resonate with the target demographic at precisely the right moment. It equips them to grasp and anticipate customer preferences, behaviours, and potential future actions.

Additionally, predictive modelling streamlines resource allocation by pinpointing the marketing channels or campaigns likely to yield the highest ROI. It enhances sales forecasting by furnishing a means for predicting sales based on historical data and market conditions. Furthermore, predictive models assist in customer segmentation and personalization, enabling marketers to provide pertinent marketing endeavours for everyone.

In a constantly evolving market landscape characterized by intense competition, predictive analytics bestows a competitive advantage by enabling businesses to identify emerging market trends and consumer behaviours before they fully emerge. This insight allows for prompt strategic adjustments and fuels innovation in product development, marketing strategies, and enhancements in customer experience.

As businesses endeavour to stay ahead in the ever-changing marketing paradigms, predictive modelling emerges as a transformative instrument that not only drives data-informed decision-making but also nurtures consumer-centric approaches. The integration of predictive analytics not only boosts marketing ROI but also heightens levels of customer satisfaction through customized experiences tailored to individual preferences [13–15].

6.2 Applications in Marketing Strategy

For organisations to effectively lead their marketing strategy, predictive modelling is an essential tool. By using advanced algorithms and machine learning, companies can analyse past data to predict future customer behaviour and lifetime value. This intelligence helps in making informed decisions, prioritizing high-value customers, and tailoring campaigns for maximum impact. Predictive analytics can also anticipate customer attrition and enable personalized retention tactics. It aids in recognizing patterns in customer behaviour and forecasting responses to new products or campaigns, ensuring alignment with customer expectations and market trends.

Moreover, predictive modelling allows for customized marketing communications by delivering personalized messages based on individual preferences. This level of personalization boosts engagement increases conversion rates and enhances ROI. By forecasting future customer acquisition, businesses can make strategic decisions supported by data-driven insights, adapting swiftly to market shifts and optimizing resource allocation for sustainable growth. In a competitive market, predicting consumer trends through predictive modelling offers an edge in crafting successful marketing strategies.

In conclusion, integrating predictive analytics empowers businesses to predict customer needs, optimize resource allocation, foster innovation, and enhance overall business agility. See references: AIContentfy [16, 17].

7 Use of Analytics for Personalized Campaigns

7.1 Importance of Personalization in Campaigns

In recent times, a fundamental aspect of strategic marketing has been the customisation of marketing campaigns to suit individual interests and behaviours. By using consumer data and analytics, companies may create offers, promotions, and messaging that are specifically tailored to the needs of their target audience. The incorporation of predictive analytics in the casino industry enables businesses to deliver customized communications to players, ultimately leading to heightened customer satisfaction and loyalty.

The essence of personalization goes beyond simply analysing data; it involves delving into customer psychology to predict, influence, and respond to consumer behaviours with finesse. Through the integration of psychological insights and marketing data, companies can develop tactics that establish a closer connection with their target audience.

Personalized marketing strategies are equally vital in the realm of business-to-business (B2B) marketing. Utilizing data-driven insights into customer behaviour allows businesses to tailor content and offers that address specific needs or pain points. In addition to increasing the relevancy of marketing materials, this degree of personalisation fosters closer bonds with prospective customers.

Data-driven decision-making plays a crucial role in lead generation and nurturing within B2B marketing as well. Marketers may optimise the performance of their acquisition tactics by scrutinising critical performance indicators like client lifetime value and conversion rates.

In summary, personalization in campaigns is no longer just a trend but a fundamental requirement in today's fiercely competitive market environment. By leveraging data analytics to understand customer preferences and behaviours, businesses can create bespoke experiences that drive engagement, foster loyalty, and ultimately fuel business growth [3, 5–7, 18, 19].

7.2 Strategies for Personalized Campaigns

To develop impactful personalized campaigns, marketers need to harness the capabilities of data analytics to customize marketing messages and offers for individual customers. Organisations can deliver targeted communications that correspond with individual consumer preferences and habits by integrating customer data into their marketing strategies. In addition to increasing client pleasure, this tailored strategy fosters loyalty, which in turn increases revenue growth.

The utilization of predictive analytics is vital for facilitating personalized campaigns by analysing extensive sets of data to predict future customer behaviour.

Understanding personal preferences allows marketers to create customised experiences that meet the unique needs of every consumer group. This practice yields improved client retention and conversion rates in addition to enhancing customer engagement.

Furthermore, the integration of AI in marketing permits real-time customization of content and offers based on ongoing customer interactions, ensuring the campaigns remain pertinent and captivating. With access to features for natural language data discovery, marketers can swiftly explore and anticipate customer preferences, leading to ultra-personalized experiences that meet customer expectations.

Businesses can create personalised and highly targeted marketing campaigns by utilising AI-driven customisation tactics and predictive data. This degree of personalisation heightens client pleasure, loyalty, and eventually revenue creation, which not only enhances the whole customer experience but also drives corporate success.

In conclusion, embedding data analytics into marketing approaches is imperative for developing personalized campaigns that drive engagement, cultivate loyalty, and optimize ROI [15, 18, 20].

8 Optimizing Acquisition Through Analytics

8.1 Acquisition Strategies

When discussing strategies for customer acquisition, optimizing marketing expenditures is key to the success of your endeavours. Utilizing customer analytics enables companies to examine data and provide critical inputs on the potential marketing channels and ways of acquiring new business relationships. This data-driven approach allows for well-informed decisions regarding where to invest resources, maximizing the return on investment while minimizing unnecessary expenses. For example, identifying high-performing channels, such as successful email marketing campaigns that lead to significant conversions, empowers businesses to redistribute funds to prioritize these channels over less effective ones like paid search advertisements. Ultimately, optimizing marketing spend through data-driven insights ensures that resources are utilized efficiently to drive customer acquisition and business growth [16].

8.2 Role of Analytics in Acquisition Optimization

Data analytics is a vital tool for improving client acquisition strategies since it offers insightful information about the behaviour and preferences of customers. By evaluating the effectiveness of current strategies, businesses can pinpoint areas for improvement and refine their approaches to attract new customers. For instance, by monitoring the performance of specific marketing campaigns, businesses can determine successful tactics and make necessary adjustments to increase their return on investment. Moreover, data analytics facilitates the identification of patterns and trends in consumer behaviour across several platforms, enabling companies to create focused marketing campaigns that appeal to their target market.

Moreover, employing predictive analytics to forecast future customer acquisition rates based on historical data can guide decisions regarding marketing budgets and resource allocation. Businesses can more successfully reach their target audience and increase engagement with their marketing efforts by utilising data-driven insights, which will ultimately result in the acquisition of more customers. Regularly monitoring and updating customer acquisition strategies with the help of data analytics is crucial for maintaining success and efficiency. By analysing customer data over time, businesses can identify trends, make well-informed decisions about strategy enhancements, and optimize their efforts to acquire new customers effectively.

In conclusion, in today's cutthroat company environment, utilising data analytics to improve and hone client acquisition methods is essential to growth and success. Through the application of customer data insights, enterprises may optimise resource allocation decisions, provide priority to critical channels, and customise customer experiences for optimal effect. Ultimately, integrating data analytics into acquisition strategies ensures alignment with business goals and drives growth efficiently [16, 17].

9 Nurturing Customer Relationships Through Data-Driven Decisions

9.1 Building Long-Term Relationships

Establishing robust and long-lasting relationships with clients is an essential part of every business plan. Businesses can obtain important insights into the needs, tastes, and behaviours of their customers by utilising predictive customer analytics. This allows them to offer targeted marketing campaigns that effectively connect with their target market. Over time, this approach fosters engagement and loyalty in addition to increasing consumer happiness.

Building long-term relationships through data-driven decisions is crucial in today's competitive business landscape [21]. Organizations are increasingly relying on data to optimize customer lifetime value, enhance customer experiences across various channels, and make strategic marketing decisions [22]. Implementing data-driven strategies not only improves the efficiency of marketing campaigns but also helps in understanding customer behaviour and expectations [23]. Moreover, in the context of plant breeding, the evolution towards Breeding 4.0 and beyond emphasizes the importance of data analytics in developing sustainable crop varieties to ensure food security [24]. By leveraging data analytics and technological advancements, organizations can foster trust, loyalty, and engagement with customers, leading to the establishment of enduring relationships based on insights and personalized experiences [25].

Predictive customer analytics empowers businesses to predict future purchasing behaviours, segment customers based on shared characteristics, and craft targeted marketing campaigns tailored to meet individual customer needs. By comprehending customer preferences and forecasting their future actions, companies can devise strategies that are more likely to connect with their audience and drive favourable outcomes.

Furthermore, customized marketing experiences result in higher customer satisfaction ratings and aid in decreasing customer turnover by identifying at-risk customers early on. By proactively implementing measures to retain these customers, businesses can nurture stronger relationships and enhance overall retention rates. Ultimately, this leads to more efficient marketing, increased conversion rates, and accelerated revenue growth.

Through the utilisation of predictive consumer analytics, companies may enhance not only their marketing initiatives but also establish long-lasting relationships with their customers. This data-centric approach enables companies to stay abreast of market trends, offer personalized experiences, and cater to the evolving demands of their target demographic. In the current competitive environment, fostering long-term relationships through predictive customer analytics is crucial for driving business triumph [26].

9.2 Utilizing Data to Enhance Customer Experience

In today's hectic business environment, building trusting relationships with clients is essential. Using customer analytics is an essential tool for learning important things about the needs, preferences, and behaviours of your customers. With this data-driven strategy, businesses can create customized experiences that meet the unique needs of every client. By delving deep into customer data, businesses can enhance customer satisfaction and foster loyalty.

Customizing customer experiences is a significant benefit of using customer analytics. Businesses can develop promotions and marketing messages that are specifically targeted to certain client segments by analysing data trends. This high level of customisation not only promotes engagement but also improves customer happiness and retention rates. Additionally, personalized experiences lead to increased conversion rates and drive overall business success.

Moreover, strategic acquisition tactics fuelled by data analysis are instrumental in elevating the overall customer experience. By studying customer data, businesses can pinpoint high-value customers and develop predictive models to anticipate their actions. This enables businesses to improve their marketing tactics and customize their messaging to meet the needs of each unique client. Consequently, businesses can enhance acquisition success rates, cut costs, and achieve growth through targeted marketing initiatives.

Businesses worldwide have been making significant investments in data analytics, and by 2023, researchers forecast a 60% rise in spending on consumer analytics. Getting all the customers' attention and keeping the competition at bay have always been challenges for the marketing department. To improve client centricity, the companies have been continuously revamping their tactics and implementing fresh methods. The advent of artificial intelligence (AI) and machine learning (ML) in the digital age is helping marketers achieve their ultimate objective of producing customized or individualized recommendations. Customer experience enhancement is limitless when unconventional data sources and methods for analysing various streams of data are available. In view of the importance and uniqueness of customer analytics, this article clarifies the sources of customer data, the tools used in customer analytics, as well as the consequences and advantages of customer analytics [27].

In essence, leveraging data to enrich the customer experience is indispensable for thriving in today's competitive market landscape. Effectively utilizing customer analytics enables companies to customize their marketing strategies, optimize acquisition tactics, and ultimately enhance overall customer satisfaction and retention rates [16].

References

1. GGI Insights.: Marketing Strategies: Craft Dynamic Approaches for Business Expansion (2024). https://www.graygroupintl.com/blog/marketing-strategies
2. McKinsey and Company.: Marketing and Sales: Big Data, Analytics, and the Future of Marketing and Sales-eBook, 1–117 (2015). https://www.mckinsey.com/~/media/McKinsey/Business%20Functions/Marketing%20and%20Sales/Our%20Insights/EBook%20Big%20data%20analytics%20and%20the%20future%20of%20marketing%20sales/Big-Data-eBook.ashx
3. Basu, R., Lim, W.M., Kumar, A., Kumar, S.: Marketing analytics: The bridge between customer psychology and marketing decision-making. Psychol. Mark. **40**(12), 2588–2611 (2023). https://doi.org/10.1002/mar.21908
4. Noha, M.: The Power of Data-Driven Decision Making: Unleashing Analytics in Strategic Marketing (2023). https://www.linkedin.com/pulse/power-data-driven-decision-making-unleashing-noha-morshed
5. Rosário, A.T., Dias, J.C.: How has data-driven marketing evolved: Challenges and opportunities with emerging technologies. Int. J. Inf. Manage. Data Insights **3**(2), 100203 (2023). https://www.sciencedirect.com/science/article/pii/S2667096823000496
6. The Importance of Data Analytics in Digital Marketing (2023). https://online.mason.wm.edu/blog/data-analytics-in-digital-marketing
7. B2B Marketing.: The Power of Data-Driven B2B Marketing—Marrina Decisions (2023). https://marrinadecisions.com/the-power-of-data-driven-b2b-marketing/
8. Vollrath, M.D., Villegas, S.G.: Avoiding digital marketing analytics myopia: revisiting the customer decision journey as a strategic marketing framework. J. Mark. Anal., 1–8 (2022). https://doi.org/10.1057/s41270-020-00098-0
9. How to design an effective customer analytics strategy for your business in 9 steps (2024). https://www.hotjar.com/customer-analytics/strategy/

10. How to Analyze and Predict the Behavior of Consumers. (2023). https://www.expressanalytics.com/blog/how-to-analyze-and-predict-the-behavior-of-consumers/
11. Asavari, T.: 5 Customer Segmentation Methods to Understand Purchasers (2022). https://amplitude.com/blog/customer-segmentation
12. Corey, P.: What is personalized marketing and how is it used today? | MarTech (2022). https://martech.org/what-is-personalized-marketing-and-how-is-it-used-today/
13. Hailey, F.: Future-Proof Your Marketing Strategy with Predictive Modeling (2024). https://improvado.io/blog/what-is-predictive-modeling
14. Roger, L.: The Power of Predictive Analytics in Shaping Marketing Strategies (2023). https://www.linkedin.com/pulse/power-predictive-analytics-shaping-marketing-strategies-roger-lopez-l1ahe
15. Woopra.: Using Predictive Analytics Marketing to Boost ROI (2023). https://www.woopra.com/blog/predictive-analytics-marketing
16. AIContentfy Team.: The benefits of utilizing customer analytics for acquisition strategies (2023). https://aicontentfy.com/en/blog/benefits-of-utilizing-customer-analytics-for-acquisition-strategies
17. AIContentfy Team.: Utilizing data analytics to inform and optimize customer acquisition strategies (2023). https://aicontentfy.com/en/blog/utilizing-data-analytics-to-inform-and-optimize-customer-acquisition-strategies
18. Elizabeth, S.: Integrating Predictive Analytics in Casino Affiliate Software for Strategic Marketing—Scaleo Blog (2024). https://www.scaleo.io/blog/integrating-predictive-analytics-in-casino-affiliate-software-for-strategic-marketing/
19. Jon, K.: Customer Analytics: How to Drive Growth from Your Insights (2023). https://www.thoughtspot.com/data-trends/analytics/customer-analytics
20. Iaroslav, G.: Key Trends in Marketing in 2024 (2024). https://www.linkedin.com/pulse/key-trends-marketing-2024-iaroslav-gumeniuk-drncf?trk=articles_directory
21. Pera, K.: Big Data for Big Decisions: Building a Data-driven Organization. Auerbach Publications (2022)
22. Tripathi, A., Bagga, T., Sharma, S., Vishnoi, S.K.: Big data-driven marketing enabled business performance: a conceptual framework of information, strategy, and customer lifetime value. In: 2021 11th International Conference on Cloud Computing, Data Science & Engineering (Confluence), pp. 315–320. IEEE (2021)
23. Stubbing, A.: Data driven decisions: a practical toolkit for library and information professionals. Facet Publishing (2022). https://doi.org/10.29085/9781783304806
24. Kuriakose, S.V., Pushker, R., Hyde, E.M.: Data-driven decisions for accelerated plant breeding. In: Accelerated Plant Breeding, Volume 1: Cereal Crops, pp. 89–119 (2020)
25. Pawlak, K.: Renewal summary report helps make data-driven decisions. Membership Manage. Report **19**(4), 4–4 (2023)
26. Susan, C.S.: Predictive Customer Analytics: Explanation + Tips (2024). https://www.pecan.ai/blog/predictive-customer-analytics/
27. Nijjer, S., Bathla, D., Sharma, S., Raj, S.: Customer analytics. In: Encyclopedia of Data Science and Machine Learning, Chap. 63, pp. 1092–1107 (2022). https://doi.org/10.4018/978-1-7998-9220-5

Leveraging Strategic Analytics to Enhance Organizational Maturity and Resilience

Amitabh Roy, Yusuf Kamal, Raj Kumar Mishra, and Saifuddin Ahmad

Abstract ***Purpose***: This chapter explores the Capability Maturity Model Integration (CMMI) as a structured approach to process maturity, resilience, and continuous improvement. It examines how organizations harness data infrastructure and their respective interpretation tools to transit from informal processes to standardized data-driven frameworks, enhancing operational efficiency, risk management, and long-term sustainability. ***Design/methodology/approach***: The study employs a comprehensive analysis of CMMI's five maturity levels through industry case studies as empirical evidence. The chapter evaluates the CMMI model's role, Meta-Data disruption, and analytics tools in process optimization and strategic adaptability. The chapter investigates companies such as Tata Consultancy Services (TCS), and Amazon and examines the benefit analysis of "Predictive Analytics", "Machine Learning" and "Artificial Intelligence". These industry references in the chapter highlight how organizations navigate maturity levels to achieve operational excellence and competitive positioning. ***Findings***: The research highlights how CMMI implementation, Meta-Data analysis and Artificial Intelligence versions leads to increased efficiency, predictable performance, and proactive risk mitigation. Organizations progressing through maturity levels achieve measurable improvements in quality management, decision-making, and overall competitiveness. The findings demonstrate that data-driven governance and continuous improvement are key drivers of business resilience. ***Originality/value***: This chapter provides a structured perspective

Theme: "Strategic Analytics and Organizational Maturity."

A. Roy
Associate Professor, Institute of Management Sciences, University of Lucknow, Lucknow, India
e-mail: roy_amitabh@lkouniv.ac.in

Y. Kamal (✉) · M. Raj Kumar · S. Ahmad
Assistant Professor, Institute of Management Sciences, University of Lucknow, Lucknow, India
e-mail: kamal_yusuf@lkouniv.ac.in

M. Raj Kumar
e-mail: mishra_rajkumar@lkouniv.ac.in

S. Ahmad
e-mail: ahmad_saifuddin@lkouniv.ac.in

S. Kautish et al. (eds.), *Strategy Analytics for Business Resilience Theories and Practices*, Information Systems Engineering and Management 33,
https://doi.org/10.1007/978-3-031-82369-5_6

on CMMI's strategic relevance in modern business landscapes and the technological engagement of organizations. By bridging theoretical frameworks with practical applications the chapter offers valuable insights for industry leaders, policymakers, and researchers. The discussion extends beyond compliance, showcasing how organizations leverage maturity models and technologies to sustain competitive advantage and innovation.

Keyword Strategic business analytics · Organizational resilience · Process maturity levels · Risk mitigation · Operational excellence · Data-driven decision making · Digital organizations · Business process optimization

1 Introduction

Organizational maturity and resilience are vital in the present business environment. Companies face unprecedented challenges which can be managed with robust strategies. However, strategic implementation alone is helpless to deliver the answers to environmental dynamics. Hence it is needed to make strategic management more unbeatable Strategic analytics provides a road to navigate these complications and answer these dynamics. Organizations are using Business Intelligence (BI) but not getting full benefits due to low maturity levels [1]. By leveraging data, organizations enhance decision-making and foster growth. Assessing Business Intelligence (BI) maturity with an impact oriented model enhances alignment with business needs and organizational performance [2].

The business landscape has dramatically shifted with due influence of digital disruptions and competition. Enterprise Social Media (ESM) significantly enhances digital maturity by transforming organizational value creation processes [3]. AI has the potential to significantly improve business resilience by enhancing risk identification, assessment, and mitigation [4]. Mature organizations manage resources and respond to changes more efficiently. While resilience ensures consistency despite disruptions. Maturity in business process management significantly enhances organizational resilience by fostering well established and adaptable processes [5]. Cross-sector partnerships within sustainable business models are vital in supporting socio-ecological resilience [6].

Besides, Globalization, technological advancements, and economic uncertainties increase the complications. Companies must innovate, adapt and control the various aspects of businesses to ensure their profitable survival. Organizational analytical capabilities and business process management maturity significantly enhance organizational resilience [7]. Organizational maturity and resilience are crucial elements. Strategic analytics is pivotal in the enhancement of Organizational Maturity and Resilience. It is based on data for visualization, implementation, evaluation and controlling the long term strategies. This approach helps organizations ensure informed decisions by improving efficiency and driving innovation. By acknowledging the strategic analytics, companies may achieve sustainable growth despite the dynamics and complexities. Entrepreneurial motivations, whether intrinsic or extrinsic, significantly shape new business performance, with extrinsic motivations favoring short-term financial gains and intrinsic motivations supporting long-term business sustainability [8].

This chapter attempts to explore, how strategic analytics elevates organizational maturity and how resilience can be powered? To achieve the objectives of this chapter, the approach is to delve into structures, practices, and real world examples, to picture the construction of organizational maturity and resilience in the world of data driven businesses. This chapter ultimately aims to provide a comprehensive handbook which covers the theoretical foundations and practical applications for enhancement of Organizational Maturity and Resilience.

2 Unveiling the Key Concepts

2.1 Demystifying Data—How Strategic Analytics Fuels Smart Decisions

In the changing world of business, having a crystal ball would be a dream come true. However, entrepreneurs do not get afraid. Strategic analytics is the next best thing. It is like sifting through a treasure trove of data, using fancy tools and techniques to unearth hidden gems that make those numbers sing. These insights are not just for show, they are the fuel that propels smart business decisions across the board.

Think of it this way. Strategic analytics is like having a GPS for your organization. It helps you see the big picture, anticipate roadblocks, and steer clear of dead ends. It not just tells you where you are, it shows you where you can go and how to get there faster and more efficiently.

But strategic analytics goes a step further than your average mapping application. Guiding a navigator about direction is not only the importance of strategic analytics, it also helps you to get familiarity about the prevailing or tentative weather. By analyzing trends and predictive modeling you may prepare for what is coming up? This proactive approach empowers you to seize opportunities before your competitors even trace them and hence you may also minimize risks before they convert into full disasters.

Therefore Strategic analytics is the bridge between data and decision making. It consumes a set of data and transforms it into a clear roadmap for success. It is about turning concrete perceptions into action and ultimately maximizing your chances of success in the business arena. Hence, pave for the assumptions and hold the power of strategic analytics. It can be used as the key for a brighter future for your organization.

Imagine building a skyscraper. Strategic analytics is like laying the foundation. It all starts with meticulous data collection, gathering the bricks and mortar of information from your internal operations and the outside world. This data gets put through the wringer with advanced tools, extracting the real gems of insight that matter most to your management team. These insights then become the blueprints for smart strategic decisions, guiding your organization towards the best course of action. But it does not stop there, strategic analytics is an ongoing process which constantly monitors and assists you in refining your approach to ensure it stays adaptable as the market shifts and your goals evolve.

In today's information age, data is not just information. It is a treasure trove waiting to be plundered. Strategic analytics acts like your personal alchemist, wielding the power to transform this raw material into a potent strategic advantage. It fuels sustainable growth, propelling your company towards unparalleled success.

By wielding this data effectively, you can fine tune your operations to a razor's edge, outmaneuvering the competition with every move. Imagine building a winning organization with the strength and endurance to dominate for years to come. Strategic analytics is your secret weapon in this dynamic business landscape. Embrace it, unlock the power of your data, and watch your business reign supreme.

2.2 The Orchestra Analogy—Understanding Organizational Maturity and Capability Maturity Model Integration (CMMI)

Imagine your company as an orchestra. Departments are instruments, each having a crucial role to play for melody. But without a conductor, the music clashes. Organizational maturity is your conductor, ensuring everyone plays in synchronization.

Higher maturity levels lead to a smoother performance, just like a well rehearsed orchestra. Each department operates seamlessly, boosting capabilities and stability. It is the difference between chaotic noise and a harmonious symphony. Your organization becomes a high-performing unit, ready to conquer any challenge.

This maturity is not one-dimensional. It is a full symphony. It covers everything from daily operations (the melody) to clear goals (the sheet music) and a data-driven approach (the conductor's keen ear). Mature organizations, like a well rehearsed orchestra, have well-defined processes, crystal-clear goals, and make decisions based on the "music" of their data. It is all about creating a harmonious and successful business environment.

Understanding the evolution and classification of organizational maturity models is crucial for aligning strategic management practices with contemporary research trends and practitioner needs [9].

In this domain, Maturity models like Capability Maturity Model Integration (CMMI) guide this progression and chronology of genesis. CMMI are frameworks used to assess and improve an organization's processes and practices. This global certification assesses organizational maturity in software design and development, with only 117 organizations worldwide currently holding it. Widely adopted in the software industry, CMMI facilitates cost reduction and minimizes rework through its structured approach [10]. CMMI defines five maturity levels. Begins from Initial (where processes are ad hoc and chaotic) up to optimizing (where continuous improvement is rooted in the organization's culture). Organizations can use these maturity models to systematically enhance their operations, to optimize resource allocation, to improve quality of products and services, and ultimately to achieve better outcomes as per their strategic objectives.

2.3 Thriving in the Business Jungle: Strategic Analytics for Resilience

In the wild jungle of business, every company needs a survival skill. That skill sounds like resilience in the present discussion. It is the ability to respond to whatever disrupts the business ecosystem. Resilience is key to stay alive and profitable in the business jungle. Strategic analytics should be used as the eyes and ears of business. They will help you spot challenges such as economic slumps or technological revolutions before they become threats to business operations and profitability.

With this early warning mechanism of your business management, you can implant strong responses that will ensure that your organization is adaptable. Many companies are already using strategic analytics like a compass through which they navigate a line of action towards innovation and industrial leadership. The companies anticipate challenges not only to respond to them but also for a strategic remedy which will ensure a healthy business. But by making resilience management a core practice, you build a strong and flexible company who is ready to face anything. Embrace strategic analytics, be proactive, and watch your business become the envy of the competition by appearing as a true champion of resilience.

Tata Consultancy Services (TCS) addresses resilience with the help of robust business continuity plans and adaptive strategies [11].

Amazon showcases resilience management with its advanced strategic analytics, promising proactive risk management as well as swift adaptation to market dynamics in e-commerce and cloud computing [12, 13].

Strategic analytics builds resilience through Sharp eyes and steady hands. It acts like a lookout, analyzing data to spot potential risks before they become problems. Imagine facing a storm, with strategic analytics, you see the dark clouds gathering and have time to prepare.

This proactive approach lets you develop contingency plans, like building a sturdy shelter before the downpour. Strategic analytics helps you mitigate risks before they are harmful, this is similar to "how car insurance protects you from unexpected losses". With real time data as your pathfinder you can constantly tailor your course of action and promise organization's agility and adaptability. In a business jungle the resilience is working as an immune system protecting your business body from viruses. Resilience also delivers the best nutrition out of your consumption. By practicing strategic analytics managers can anticipate threats and gain business excellence.

3 Interplay Between Strategic Analytics and Maturity Models

Imagine a ballroom, a stage for performance. One dancer represents strategic analytics while the other is an organizational maturity model. The steps of dancers are calculated and deliberate. Although both the dancers have different body movements but ensure a complimentary node for the other. The synergy produces grace in the performance. Strategic analytics guide the organization towards excellent performance. While organizational maturity models provide a foundation for stability.

In an organization, strategic analytics drives the dance through the directions to collect data and the music of the business is the analysis of data to reveal hidden truth. The guiding hand on the dance floor for the strategic analytics is organizational maturity model. It supports its partner as an instructor to assess the organization's current capabilities.

It identifies areas for improvement, opportunities to refine their steps. Based on this assessment, the maturity model recommends a series of moves. These might involve data governance strategies, talent development programs, or the implementation of specific analytical tools. With each step, the organization builds its skills, mastering the dance floor.

The interplay between these two creates a grace production cycle. Strategic analytics is fueled by the insights gleaned from data and informs the maturity model's recommendations. The maturity model in turn strengthens the organization's analytical capabilities enabling it to gather even richer data and to generate even more profound insights. It is a continuous loop of improvement.

This synergy is not just about grace and elegance. It is about achieving tangible results. As the organization matures its analytical capabilities it becomes more proficient at making data driven decisions. This translates into improved efficiency, better risk management, and a stronger competitive advantage. The organization becomes a champion on the dance floor of business and then is able to navigate the complexities of the market with confidence and agility.

4 Components of Strategic Analytics Implementation

4.1 Building the Bedrock: Data Infrastructure for Strategic Analytics

Imagine a magnificent skyscraper, a testament to human ingenuity. But what most people do not see is the foundation beneath, the intricate network of steel and concrete that ensures its stability. Strategic analytics is also a powerful tool, however its efficiency is directly proportional to the data infrastructure.

This data infrastructure is that incubation center which collects, preserves, and manages the data in alignment with strategic analytics.

4.1.1 Data Warehousing

"Data Warehousing" can be seen as a storage facility for data. The data warehousing works as a library for strategic analysts, this library provides data from all the functions and departments of an organization to the strategic analyst. The data warehouse is an option for easy access and retrieval of data for a strategic analyst. The data warehouse shall have the cataloging kind of mechanism to cater data needed.

4.1.2 Data Lakes

Unlike the structured slots of a library and data warehousing, the "Data Lakes" are more vast and unrefined reservoirs of data. The "Data Lakes" store all types of data in its raw form to create a rich pool of data for a business organization.

4.1.3 Data Integration Tools (DIT)

Data Integration Tools implies those systems which are used to bridge among. These DITs are used to "Extract", "Transform" and "Load" (ETI) the data so that the data can be converted into a usable version for integration after elimination of inconsistencies in the data. This ETI (Extract Transform Load) is an important principle that can be used by various organizations as a foundation for development of their own DIT set up.

4.1.4 Data Governance

Data is a powerful asset however in absence of proper supervision and handling it cannot be harnessed to accomplish the organizational goals. The "Data Governance" looks after the framing and implementation of policies and procedures to ensure the quality and security. It can be seen as a "Rule Book" of the library to provide easy and clean access to the right data needed.

Organizations are supposed to continuously invest in "Data Infrastructure" in the dynamic business environment. A sound "Data Infrastructure" foundation enhances the benefits of strategic analytics as well as for navigation of dynamic business landscape.

4.2 *Equipping the Workforce: The Analytics Tool Belt*

Workforce empowerment with the right tool kit is a traditional rule of business. Any employee at any level in an organization needs to rely on the right set of tools and techniques to use those skills. This is not only what an organization needs to harness the potential of an employee, besides, identification of the right employee is its integral part.

4.2.1 Data Visualization Tools

These are the paint brushes and canvases of the data world, transforming raw numbers into compelling charts, graphs, and dashboards. They help bring data to life and make complex insights readily digestible for everyone in the organization.

4.2.2 Data Ninjas: BI Tools Slice and Dice to Success

Are you drowning in data? BI tools are your secret weapon. Imagine them as ninja tools for company info. Users can analyze, uncover trends, and answer questions on the basis of real data not hunches. What is the best part? Everyone can be a data ninja. BI tools empower all to make data driven decisions. Ditch the guesswork, unlock insights, and fuel success.

4.2.3 Data Gurus—Beyond BI with Advanced Tools

BI tools are cutting edge technologies, however they differ in their utilities and deliverables. Tools like predictive analytics forecast trends while machine learning uncovers hidden patterns, and the Natural Language Processing unlocks text insights.

Fancy features of the BI may attract the users however choosing the right tool will build the immune system of business. Identify the need in the business, upgrade team skills, and prioritize user friendly and scalable options during the selection of BI tools.

Empower the team with the right "data detection" tools. It becomes your responsibility to unearth the hidden gems and translate insights into action that is the key to success.

4.3 The Human Edge: Talent and Skills for Strategic Analytics

The most sophisticated data infrastructure and cutting edge analytics tools are merely inert components without the human element. The skilled individuals who are passionate into strategic analytics shall be the part of the team.

Technology is a booster, but it cannot replace humans and in fact need them to deliver the results. Data is powerless without human involvement during the interpretation.

4.3.1 Understanding the Business Context

Data does not have throat to produce voice for communication. It is the analyst who puts a voice producing machine in the data after which only the data becomes communicative. Insights' translation into an understandable language is the key exercise taking place.

4.3.2 Asking the Right Questions

Data is like an ocean. It is you who might get lost without the right captain. Data Analysts are like curious navigators, who ask smart questions to unveil the most relevant insights. Data Analysts frame a course to success and not just swim aimlessly.

4.3.3 Critical Thinking and Problem-Solving

Data can sometimes be messy and ambiguous. Analysts need to be critical thinkers, able to sift through conflicting information, identify patterns, and draw sound conclusions. They are the data detectives, piecing together the puzzle to solve complex business problems.

Building a skilled analytics team requires a multi-pronged approach, hence we recommend.

Bridge the Skills Gap

Not everyone comes equipped with the necessary skill set. Invest in training programs to equip your existing workforce with the essential analytics skills, such as data analysis, data visualization, and critical thinking.

Invest in Talent Acquisition

Seek out individuals with a blend of analytical skills, business acumen, and strong communication abilities. Look for those who are passionate about data and eager to translate insights into action.

Foster a Culture of Learning

Data is a constantly evolving field. Encourage a culture of continuous learning within your organization, providing opportunities for your team to stay abreast of the latest trends and technologies in the analytics domain.

Remember, your people are your greatest asset in the realm of strategic analytics. By nurturing their talent, fostering a culture of data-driven decision making, and providing them with the right tools, you empower them to become the architects of your organization's data-driven future.

5 Cultivating a Thriving Analytics Ecosystem: Best Practices for Success

Strategic analytics is not a solitary endeavor; it flourishes within a well-nurtured ecosystem. This ecosystem is the fertile ground where data, talent, technology, and culture converge to empower data-driven decision making. Here are some best practices to cultivate a thriving analytics ecosystem:

5.1 Executive Buy-In is the Cornerstone

Building a robust analytics ecosystem requires leadership from the top. Executives need to champion the cause, allocating resources, fostering a data-driven culture, and setting clear expectations for data utilization. Without their support, strategic analytics becomes a mere buzzword, gathering dust on the shelf.

5.2 Align Analytics with Business Goals

Do not get lost in the labyrinth of data for data's sake. Ensure your analytics initiatives are closely aligned with the organization's overall strategic goals. Focus on collecting and analyzing data that directly addresses key business challenges and informs strategic decision making.

5.3 *Democratize Access to Data*

Break down the data silos. Empower a wider range of employees to access and analyze data relevant to their roles. This fosters a culture of ownership and accountability, where everyone feels empowered to leverage data for continuous improvement.

5.4 *Break Down Communication Barriers*

Data scientists and business users often speak different languages. Bridge this gap by fostering effective communication channels. Data scientists need to be able to translate complex insights into actionable recommendations that resonate with business leaders.

5.5 *Embrace a Culture of Experimentation*

Do not be afraid to experiment and learn from failures. Encourage a culture of innovation, where teams can test new analytics tools and approaches to see what works best for your organization. Remember, the most successful journeys often involve a few wrong turns along the way.

5.6 *Invest in Data Governance*

Data is a powerful asset, but like any valuable resource, it needs proper management. Implement a robust data governance framework to ensure data quality, security, and accessibility. Think of it as setting the ground rules for using data within your organization, ensuring everyone has access to the right information at the right time.

By nurturing these best practices, you cultivate a thriving analytics ecosystem and a fertile ground where data blossoms into actionable insights that propel your organization towards a data driven future. Remember, a robust ecosystem is not built overnight; it is a continuous journey of refinement and improvement.

6 From Theory to Practice

So, we have delved into the theoretical underpinnings of strategic analytics and organizational maturity. Now, let us roll up our sleeves and explore how these concepts translate into real world action. Buckle up, because we are about to embark on a journey through the annals of strategic analytics success stories, case studies that showcase the transformative power of data-driven decision making.

Remember that marketing campaign you once designed for a local coffee shop, the one that targeted busy professionals with laser focus? Well, strategic analytics operates on a similar principle, only on a much grander scale. It is about understanding your target audience. And in this chapter it is the intricate working of your organization and leverage data to make informed choices that drive growth and competitive advantage.

These case studies look like adventurous trips of business. We will talk about those that have successfully cultivated the power of strategic analytics. We will also investigate how they identify the significance of their data. Our discussion will include the analysis of "How to craft a robust analytics structure?" and "how to sketch data driven insights for achieving remarkable results?" It can be a retailer who harnesses predictive analytics to optimize the Supply Chain Management, or it can be a healthcare service provider who leverages the potential of data for improving patient's health.

These cases portray two dimensions. First they demonstrate concrete examples of how strategic analytics translates into tangible benefits? Second, they offer valuable lessons on what are the pitfalls and how to avoid them and the best practices to emulate. Treat cases as guides for your journey towards organizational excellence through strategic analytics.

7 Benefits of Strategic Analytics for Organizational Maturity

Imagine your organization as a magnificent tree house, a testament to ingenuity and collaboration. But just like any tree house worth its salt, it needs a solid foundation and well placed beams to reach for the sky. Strategic analytics acts as the master carpenter in this scenario, meticulously crafting the framework that empowers your organization to climb the ladder of maturity.

Now, picture yourself perched in this very tree house, gazing out at the sprawling landscape of the business world. Strategic analytics equips you with a high powered telescope, allowing you to see opportunities and challenges with unparalleled clarity. You can anticipate shifts in market trends, identify potential roadblocks before they arise, and make informed decisions that propel your organization towards a brighter future.

But the benefits extend beyond mere foresight. Strategic analytics is like a Swiss Army knife for organizational improvement. It equips you with a variety of tools. Below are the benefits listed.

7.1 Data-Driven Decision Making

No more flying blind, Strategic analytics replaces guesswork with concrete evidence, allowing you to make strategic choices based on real data, not hunches.

7.2 Enhanced Efficiency

Imagine streamlining processes, eliminating wasteful practices, and optimizing resource allocation. Strategic analytics helps you identify areas for improvement and implement data driven solutions that tighten your organization's belt.

7.3 Improved Risk Management

The business world is full of unforeseen storms. Strategic analytics acts as your early warning system, helping you anticipate potential risks and develop contingency plans to weather any storm.

7.4 A Culture of Continuous Improvement

Strategic analytics does not stop at a single success. It fosters a culture of data-driven inquiry, where learning and improvement become a continuous journey.

The rewards of strategic analytics are as bountiful as the fruits growing on the branches of your metaphorical treehouse. It empowers your organization to not just survive, but thrive, in the ever-evolving landscape of the business world. So, grab your metaphorical hammer and nails, and let us embark on the rewarding journey of building a stronger, more mature organization with the help of strategic analytics.

8 Beyond the Tool Box: Advanced Analytics Techniques

So far, we have discussed the foundational tools and methodologies for strategic analytics. But for those seeking to delve deeper, a treasure trove of advanced techniques awaits. These are the specialized power tools that unlock even richer insights from your data, akin to wielding a laser instead of a screwdriver.

Here, we will take a peek into three of these advanced approaches:

8.1 *Predictive Analytics: Seeing Around the Corner*

Imagine yourself as a seasoned explorer, navigating a dense jungle. Predictive analytics is your trusty map, one that not only shows your current location but also predicts the path ahead. By analyzing historical data and identifying patterns, predictive analytics helps you forecast future trends, anticipate customer behavior, and make informed decisions about resource allocation and marketing strategies.

Think of it like predicting the weather. By analyzing historical data on temperature, humidity, and wind patterns, meteorologists can forecast the likelihood of rain or sunshine. Similarly, predictive analytics allows you to make educated guesses about what is coming down the pike in your industry.

8.2 *Machine Learning: Unveiling Hidden Gems*

Ever get that uncanny feeling that your favorite online retailer can read your mind? That is the power of machine learning at play. These algorithms are like self learning apprentices, constantly improving their ability to identify patterns and relationships within data. The more data they are fed, the smarter they become, uncovering hidden gems and insights that traditional analytical methods might miss.

Imagine sifting through a mountain of customer reviews to understand what truly resonates with your target audience. Machine learning can automate this process, identifying key themes and sentiment within the data, saving you valuable time and effort.

8.3 *Artificial Intelligence: The Data Whisperer*

Artificial intelligence (AI) takes machine learning a step further. Think of it as the ultimate data guru, capable of not just identifying patterns but also extracting meaning and context. AI algorithms can analyze vast amounts of complex data, including text, images, and even speech, to generate insights that would be impossible for humans to glean on their own.

For instance, imagine analyzing social media conversations to understand customer sentiment towards your brand. AI can process the nuances of language, sarcasm, and even emojis to provide a more comprehensive picture of how your customers perceive your organization.

These are just a taste of the advanced analytics techniques at your disposal. While they offer immense potential, it is important to remember that they are not magic bullets. Success hinges on having a clear understanding of your business goals, choosing the right tools for the job, and ensuring you have the data quality and infrastructure to support these sophisticated approaches.

9 Navigating the Bumpy Road: Challenges in Implementation

The path to strategic analytics nirvana is not always smooth sailing. There will be bumps along the road, roadblocks to overcome. But fear not, intrepid explorer. Forewarned is forearmed. Let us tackle some of the most common challenges you might encounter on your journey.

9.1 Data Quality Issues: Garbage in, Garbage Out

Imagine building a magnificent sandcastle on the beach, only to have the tide wash it away. That is what happens when you build your strategic analytics efforts on a foundation of poor quality data. Inconsistent formats, missing values, and inaccuracies can all lead to misleading insights and ultimately, failed initiatives.

The key is to prioritize data quality from the get-go. Establish clear data governance policies, invest in data cleansing tools, and foster a culture of data accuracy within your organization. Remember, your data is the fuel that powers your analytics engine hence keep it clean and well-maintained.

9.2 Data Deluge: Navigating the Resistance

Change, even the data driven functioning, can be a bumpy ride. Some folks might feel their expertise or control is on the line. Others might just dislike new systems.

So what is the key to overcoming this resistance? Clear communication and a solid change management plan. Demonstrate how strategic analytics benefits everyone? Organize proper training and support to help everyone navigate this data revolution.

9.3 Building Your Analytics Team

Practicing strategic analytics is a team activity similar to any championship or contest. You need the right players on the field to ensure victory. This means identifying skill gaps within your workforce and correspondingly investing in training programs to ensure a skilled team for data analysis, data visualization, and critical thinking. Blend of analytical expertise and sound business acumen might be hot cake in a bakery.

Although the creation of a skilled analytics team is a repetitive exercise, which is done through continuous monitoring and evaluation. Suggestion is to stay updated with latest trends and technologies in the industry, and promote a culture of continuous learning. You have to invest in the future through investment in your workforce.

10 Charting a Course: Strategies to Overcome Challenges

The roadblocks we discussed in the previous section might seem daunting, but do not get afraid. There are effective strategies to navigate these challenges and ensure your strategic analytics journey reaches its successful destination. Let us now explore some methodologies to tackle these barriers.

10.1 Data Defense: Guarding Against the Garbage in, Garbage Out Trap

Strategic analytics is only as good as the data it uses. Imagine building a fortress to defend your kingdom, but the walls are made of sand. That is what happens with dirty data as it weakens your defenses and leads to unreliable results. Here is your action plan to build a fortress of clean data:

10.1.1 Data Shield: Keeping Your Information Clean and Secure

Imagine building a house on a foundation of sand. Not the most secure, right? Clean, high quality data is the bedrock of strategic analytics, your foundation for success. That is why you need a data shield to keep your information clean and secure. This means establishing clear rules for how data is collected, stored, and accessed. These procedures act like blueprints, ensuring consistency and minimizing errors. Think of it as preventing "garbage in, garbage out" situations. With a strong data shield, you can build reliable insights that fuel your strategic decisions.

10.1.2 Data Scrubbers—Wiping Away the Grime

Clean data is the key to unlocking powerful insights. That is where data scrubbers come in. Imagine them as a team of janitors meticulously cleaning your information. These tools automate the process of identifying and fixing inconsistencies, missing values, and formatting errors. Think of them as wiping away the grime to ensure your data shines. With clean information as the foundation, your analytics can produce sparkling insights that illuminate the path to success.

10.1.3 Data Detectives: Everyone on Watch for Accuracy

Clean data starts with a team effort. Cultivate a "data-centric" mindset across the organization. Think of data as the lifeblood of your analytics engine. Train employees on why data quality matters and empower them to be data detectives. This means encouraging them to be vigilant, identify potential inconsistencies, and report any issues they find. With everyone on the lookout for errors, you can ensure the accuracy of the information that fuels your analytics and propels your decisions forward. Remember, clean data is the foundation for building reliable insights and achieving success.

Data quality is not a one-time fix, it is a detective's ongoing investigation. Regularly monitor your data stream for inconsistencies, becoming data detectives on the lookout for clues of potential issues. Be proactive in addressing any problems you find. Remember, a clean and reliable data stream is the key to unlocking actionable insights that fuel smart decisions. Think of it as keeping the data pipeline flowing smoothly so your analytics engine can generate accurate and powerful results.

10.2 Change Management: Guiding Your Team Through the Data Transformation

Change, as they say, is inevitable. Here is how to ensure your team embraces the data revolution.

10.2.1 Communicate the Benefits

Clearly articulate how strategic analytics will benefit the organization and individual employees. Highlight how data driven decision making can lead to improved performance, increased efficiency, and greater job satisfaction.

10.2.2 Embrace Transparency

Be open and honest about the changes involved in implementing strategic analytics. Address concerns and provide opportunities for feedback throughout the process.

10.2.3 Invest in Training and Support

Ensure the organizational supplies needed in the navigation of the data landscape and graduate your team with the skills and knowledge. We prescribe training programs on data analysis on data visualization which will let the team learn to use new technologies.

By supporting the culture of open communication and assistance you can manage resistance during the transformation of business. Also you need to demonstrate the value of data driven decision making.

10.3 Skill Development and Training: Building a Competent Data Crew

The Strategic analytics team is working as an asset. Here is the recipe to cultivate their skills.

10.3.1 Identify Skill Gaps

We would recommend a perfect assessment of your team's current analytical capabilities and identification of opportunity areas where training is needed. The bridging them will create a perfect match to the match between data and strategic visualization.

10.3.2 Invest in Training Programs

Ensure chances to develop essential skills like data analysis, data visualization, and critical thinking. Develop a combo of in house training, online, and professional development workshops.

10.3.3 Encourage Continuous Learning

The field of data analytics is constantly evolving. Foster a culture of continuous learning within your organization, encouraging employees to stay abreast of the latest trends and technologies.

Remember, building a skilled analytics team is a marathon, not a sprint. Invest in your people, provide them with the resources they need to succeed, and watch your team evolve into a well oiled data analysis machine.

11 Crystal Ball Gazing: Future Trends

The world of strategic analytics is a dynamic landscape, constantly evolving with the emergence of new technologies and methodologies. So, fasten your seatbelts, fellow data adventurers, as we take a glimpse into the future and explore some of the exciting trends that are poised to reshape the way organizations leverage data.

11.1 Emerging Technologies: A Glimpse into the Future's Analytics Toolbox

Get ready to witness an ever-expanding arsenal of tools at your disposal. Here are a few emerging technologies that promise to revolutionize strategic analytics.

11.1.1 Natural Language Processing (NLP)

Imagine a world where computers can understand human language as fluently as we do. NLP advancements will allow you to extract insights from vast troves of unstructured data, like text documents, social media conversations, and customer reviews.

11.1.2 Quantum Computing

Think of it as a supercharged computer on steroids. Quantum computing holds the potential to tackle complex analytics problems that are currently beyond the reach of traditional computers, ushering in a new era of data discovery.

11.1.3 Edge Computing

Data is no longer confined to centralized servers. Edge computing brings processing power closer to the source of data, enabling real-time analytics and faster decision making.

These are just a taste of the technological advancements that are on the horizon. The key is to stay informed, identify the technologies that align with your organizational goals, and be prepared to adapt your approach as the landscape continues to evolve.

11.2 Methodologies: Refining the Art and Science of Data Analysis

The "how" of strategic analytics is constantly being refined? Here are some emerging methodologies that are worth keeping an eye on.

11.2.1 AI in the Spotlight—Lifting the Lid on Black Boxes

Imagine you are using a powerful new tool, but you are not sure exactly how it works. That is kind of what AI can be like in analytics. XAI, or Explainable Artificial Intelligence, methods act like floodlights, illuminating the inner workings of AI and how it reaches its conclusions. This transparency builds trust and encourages wider adoption of this powerful technology.

11.2.2 Data on Demand—Making Decisions in the Fast Lane

Imagine having access to insights the moment the data rolls in. Real-time analytics is like having a live feed of information, empowering you to make decisions on the fly. This lets you seize fleeting opportunities and dodge potential risks before they become problems. It is all about capitalizing on the "now" of data for faster, smarter choices.

11.2.3 Data Fortune Tellers—Predicting the "What if"

Data can describe the past and predict the future, but prescriptive analytics goes a step further. Think of it as a data fortune teller, peering into "what if" scenarios. These techniques use historical data and real time insights to recommend the best course of action, empowering you to proactively shape your organization's future. It is not just about seeing what is coming, it is about using data to take control of your destiny.

The world of data is constantly evolving. By staying on top of these cutting edge methodologies, you ensure your strategic analytics stay sharp and effective for years to come. Think of it as keeping your analytical toolbox well stocked with the latest and greatest tools.

11.3 Continuous Learning and Adaptation: The Cornerstone of Future Success

The most important trend of all? A relentless commitment to learning and adaptation. Strategic analytics is an ongoing expedition, not a treasure hunt with a fixed endpoint. Here is your roadmap to stay ahead of the curve.

11.3.1 Data Detectives—Unleashing the Power of Curiosity

In the data jungle, curiosity is your compass. Encourage your team to be data detectives, asking questions, exploring uncharted territories, and embracing experimentation. Remember, a curious mind is a goldmine in the ever-evolving world of data.

11.3.2 Invest in Upskilling

The data landscape is ever changing. To stay ahead, equip your data ninjas with the latest skills. Invest in ongoing training to keep your team at the forefront of the data analytics revolution. It is about ensuring they have the tools they need to leverage powerful new technologies and methodologies effectively.

11.3.3 Embrace Agility

The world of data is a constant evolution. The winning teams will be those who can adapt like chameleons, embracing new tools and techniques as they emerge. Be prepared to adjust your approach, and prioritize continuous improvement of your strategic analytics practices. Remember, agility is key to unlocking the full potential of data and staying ahead of the curve.

The world of data is a vast ocean, brimming with potential. But to truly thrive, you need to be a curious crew, forever learning and innovating. Embrace change like a steady wind at your back, continuously refining your approach to strategic analytics.

Remember, in this data-driven odyssey, the future belongs to those who can harness the power of information. Make informed decisions, navigate challenges with agility, and adapt your course as needed. Embrace the unknown, for within it lie exciting opportunities.

So, set sail on your data voyage, captain your team of data detectives, and conquer the uncharted territories of analytics. The age of information is yours to explore, and success awaits those who chart their course wisely.

12 Conclusion: The Power of Data-Driven Decisions in Organizational Maturity

As we reach the conclusion of this exploration, let us take a moment to reflect on the transformative potential of strategic analytics. We have delved into the foundational elements, like the data infrastructure, the analytical tools, and the skilled personnel which form the bedrock of a robust analytics ecosystem. We have navigated the challenges that can arise on this journey, from data quality issues to skill gaps, and explored strategies to overcome them. But most importantly, we have peered into the future, witnessing the exciting new technologies and methodologies that promise to reshape the field of strategic analytics.

Throughout this chapter, we have endeavored to share this knowledge not just as a professor, but also as a fellow traveler on the path to data driven maturity. Remember that captivating story about the struggling coffee shop marketing campaign I mentioned earlier? Well, that story has a happy ending. By leveraging strategic analytics, the coffee shop was able to identify its target audience with laser focus, personalize its marketing messages, and ultimately achieve significant growth. This is just one example of the transformative power of data driven decision making.

As you embark on your own strategic analytics journey, remember this. The data is a powerful tool, but it is only as valuable as the insights you can extract from it. By building a robust analytics ecosystem, investing in your people, and embracing a culture of continuous learning, you empower your organization to unlock the hidden potential within its data. This, in turn, fuels organizational maturity, enabling you to navigate the ever-changing business landscape with agility, make informed decisions that drive growth, and ultimately, achieve a sustainable competitive advantage.

The future belongs to organizations that can harness the power of data. So, leverage the tools and strategies explored in this chapter, cultivate a data-driven mindset within your organization, and embark on your own journey of data-driven transformation. Remember, the most rewarding discoveries often lie just beyond the horizon, waiting to be unearthed by the curious and data-savvy explorer.

References

1. Ong, I.L., Siew, P.H., Wong, S.F.: Assessing organizational business intelligence maturity. In: ICIMU 2011: Proceedings of the 5th international Conference on Information Technology & Multimedia, Kuala Lumpur, Malaysia, pp. 1–6 (2011). https://doi.org/10.1109/ICIMU.2011.6122732
2. Lahrmann, G., Marx, F., Winter, R.: Business intelligence maturity: development and evaluation of a theoretical model. In: 2011 44th Hawaii International Conference on System Sciences, Kauai, HI, USA, 2011, pp. 1–10 (2011). https://doi.org/10.1109/HICSS.2011.90
3. Feitosa Jorge, L., Mosconi, E., Santa-Eulalia, L.A.d.: Enterprise social media to foster digital maturity: a value-creation perspective. J. Syst. Inf. Technol. **26**(3) (2024). https://doi.org/10.1108/JSIT-11-2023-0259

4. Ramachandran, K.K., Vinjamuri, L.P., Ramesh, R., Al-Taee, M., Alazzam, M.B.: Using AI for risk management and improved business resilience. In: 2023 3rd International Conference on Advance Computing and Innovative Technologies in Engineering (ICACITE), Greater Noida, India, 2023, pp. 978–982 (2023). https://doi.org/10.1109/ICACITE57410.2023.10182662
5. Sincorá, L.A., Oliveira, M.P.V.d., Zanquetto-Filho, H., Alvarenga, M.Z.: Developing organizational resilience from business process management maturity. Innov. Manage. Rev. **20**(2), 147–161 (2023). https://doi.org/10.1108/INMR-11-2021-0219
6. Dentoni, D., Pinkse, J., Lubberink, R.: Linking sustainable business models to socio-ecological resilience through cross-sector partnerships: a complex adaptive systems view. Bus. Soc. **60**(5), 1216–1252 (2021). https://doi.org/10.1177/0007650320935015
7. Sincorá, L.A., Oliveira, M.P.V.d., Zanquetto-Filho, H., Ladeira, M.B.: Business analytics leveraging resilience in organizational processes. RAUSP Manage. J. **53**(3), 385–403 (2018). https://doi.org/10.1108/RAUSP-04-2018-002
8. Debrulle, J., Maes, J., Gardiner, E.: How entrepreneurs think: financial decisions for the long or short term. J. Bus. Strateg.Strateg. **42**(4), 233–239 (2021). https://doi.org/10.1108/JBS-04-2020-0068
9. Kucińska-Landwójtowicz, A., Czabak-Górska, I.D., Domingues, P., Sampaio, P., Ferradaz de Carvalho, C.: Organizational maturity models: the leading research fields and opportunities for further studies. Int. J. Qual. Reliab. Manage. **41**(1), 60–83 (2024). https://doi.org/10.1108/IJQRM-12-2022-0360
10. Morales-Fernández, R., Brito-Rojas, J.A., Villaseñor Marcial, J.A.: Capability maturity model integration (CMMI). J. Res. Inform. Technol. **2**(3), 15–20 (2014). https://doi.org/10.36825/RITI.02.03.003
11. TATA Consultancy Services.: Integrated Annual Report [2022–23]. Innovate, Adapt, Thrive. TCS (2023). Retrieved 1 July 2024 from https://www.tcs.com/content/dam/tcs/investor-relations/financial-statements/2022-23/ar/annual-report-2022-2023.pdf
12. Ionescu, S.-A., Diaconita, V.: Transforming financial decision-making: the interplay of AI, cloud computing and advanced data management technologies. Int. J. Comp. Commun. Control **18** (2023). https://doi.org/10.15837/ijccc.2023.6.5735
13. Tsang, Y.P., Fan, Y., Lee, C.K.M., Lau, H.C.W.: Blockchain sharding for e-commerce supply chain performance analytics towards Industry 5.0. Enterprise Inf. Syst. **18**(4) (2024). https://doi.org/10.1080/17517575.2024.2311807

A Bibliometric Approach to Success Indicators of Entrepreneurship in Business Incubators and STEM Start-Ups

Rosy Dhall, Mikul, and Pranav Saraswat

Abstract Current study aims to precisely define the function played by technology business incubators in fostering the sustainability and expansion of technology businesses. Additionally, we made an effort to provide a thorough description of technology entrepreneurship that included references to recent advances in science and shifting fashion trends in the digital age. With increasing usage and reliance on digital technologies, it is essential for start-ups to keep up with these developments and adapt to them in order to survive and compete. Here, we are looking into the effects of the link between innovation in science and technology and business creation. This article examines how a rise in the number of support centres can affect the ways in which companies create and implement new technologies.

Keywords Business · Entrepreneurship · Technology · Start-up · Incubator

1 Introduction

Using and adapting to new technologies is crucial if one is to maintain a competitive edge in today's globalised world marketplaces. Companies are expected to use and embrace research and development processes as part of a number of science and technology policies aimed at increasing their competitiveness. Science, technology, entrepreneurship, and innovation policies are typically implemented independently

R. Dhall
Gandhinagar University, Kalol, India
e-mail: rosy.dhall@gandhinagaruni.ac.in

Mikul
Maharshi Markandeshwar University, Ambala, India
e-mail: mikul@mmumullana.org

P. Saraswat (✉)
Nirma University, Ahmedabad, India
e-mail: drpranav.saraswat@nirmauni.ac.in

S. Kautish et al. (eds.), *Strategy Analytics for Business Resilience Theories and Practices*, Information Systems Engineering and Management 33,
https://doi.org/10.1007/978-3-031-82369-5_7

of one another. To develop interventions that address technological entrepreneurship specifically, it is necessary to consider the interconnections between these factors. Entrepreneurial deliberation is influenced by the high cost of developing and deploying new technologies. In general, the more cutting-edge a technology is, the more dangerous its results will be. However, it also results in a significant financial gain. Business owners who want to successfully implement new technologies need ready access to relevant data and study areas. Early-stage entrepreneurs are attempting to develop and deploy solutions to tackle specific client problems with the help of funding agencies, thanks to the proliferation of start-up promotion schemes. During this time, the start-up's other stakeholders must provide more support for the entrepreneur to reduce the likelihood that the business would fail. Incubators for start-ups in tech are one example. Technology start-ups need more than just money; they also need direction, the ability to gauge market demand, assistance in organising the rollout of their technology, etc. Institutional frameworks including as incubators, accelerators, and mentor networks are tasked with fulfilling these roles.

The phenomena that science inspires technological innovation is similarly obvious, as is the fact that science provides insight and evidence to address challenges that may arise during the R&D process [1, 2]. Yet, many times the scientifically based innovations that emerge from research are either too early or too late to meet the needs of the market as it is now structured [3]. Therefore, it may be more dangerous to perform scientific studies via entrepreneurial means because of the greater involvement of scientific information and research to these studies [4]. Understanding the interplay between scientific inquiry and technology application is crucial for sustaining a positive effect [5]. When viewed from the perspective of information, the idea of combination can be defined as the process of combining scientific and technological data to create an end result that has not just a singular value but also the benefits of shared, unified data [6, 7]. The integration process yields an invention leading to the development of a product with higher potential than methods developed exclusively for science or technology due to the respective contributions and impacts of these two fields [8].

The concept of heroic action was derived from individuals' risky engagement in the economy between 1815 and 1885. Ability to raise finance and manage production factors required strategic foresight, leadership, awareness, and an organised approach as part of such entrepreneurial activity. According to popular belief, J. S. Mill was the first to use the term "entrepreneur" in the field of economics in 1898. Mill used the phrase to differentiate between the roles of capitalist and manager by emphasising the risk-taking nature of entrepreneurs. On the other hand, Frank [9] introduced the idea that the entrepreneur/founder is responsible for weathering the storm of fluctuating market conditions. These are necessary to handle functional expertise, such as directed and controlled structures. Entrepreneurship, as examined by [10] in light of the necessity of innovation inside a business, is shown to be a "creative destruction method of development." According to Schumpeter, those whose job it is to only arrange, and compile resources can be called "managers" of new businesses, but they cannot be considered "entrepreneurs." He makes it apparent that originality is what truly sets entrepreneurs apart.

If a company is innovative and entrepreneurial, it will find a way to add value to an existing resource by discovering a previously unrecognised way to put it to use. By his own words, Schumpeter implies:

> An innovation could take place in product design, organization of the firm, marketing devices or process design. Innovation is what separates the entrepreneur from others who undertake closely related endeavours. Entrepreneurship is crucial for understanding economic development.

Objectives of the Study:

(A) To provide a working definition of science and technology entrepreneurship.
(B) Goal: to evaluate the long-term viability and performance of India's tech-based start-ups.
(C) To examine how technology business incubators contribute to the development of a vibrant start- up and entrepreneurship scene.
(D) Using quantitative and qualitative methods, to present the conceptual relationship between science and technology, technology business incubators, technology start-ups, and the entrepreneurship ecosystem.

2 Literature Review

The nature of entrepreneurship in relation to founder qualities, growth, macroeconomic perspective, etc., has been the subject of numerous attempts at in-depth analysis [11–17]. There have been a number of additional attempts to examine the growth of entrepreneurship in developing nations from a strictly macroeconomic perspective. It's safe to say that most emerging nations, including India, spent the majority of the twentieth century in a period of transition. Just the development of a diverse group of self-starters made this shift possible. To achieve this goal, it was necessary to expand the pool of people capable of engaging in entrepreneurial activities. Entrepreneurial activity grew rapidly in cities at first. Since the beginning of the Commercial Revolution, businesses and entrepreneurs have played a pivotal role in driving modernization forward. This concept has been investigated by economists and sociologists, primarily at the boundaries of their respective fields.

Entrepreneurship, as defined by the National Knowledge Commission (NKC) in 2008, is "the professional application of knowledge, skills and competencies and/or of monetizing a new idea, by an individual or a set of people by launching an enterprise de novo or diversifying from an existing one" (as opposed to "seeking self-employment as in a profession or trade"). It highlights "Individual Motivations, Socio-Cultural Elements, Access to Early-Stage Finance Education and Business Environment" as four essential factors that encourage entrepreneurship (Fig. 1).

The rate of innovation among business owners is highest in India (47%), according to the 2018–19 edition of the Global Entrepreneurship Monitor (GEM) report. Similarly, the percentage of innovative entrepreneurs in India is high despite the country's low TEA. According to the GEM-created National Entrepreneurship Context Index

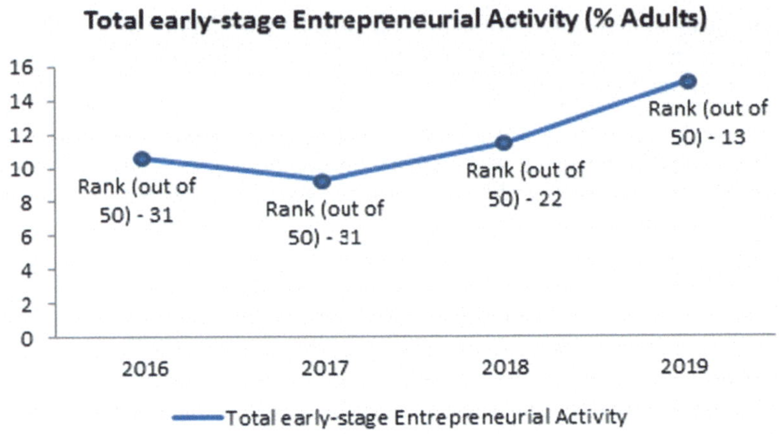

Fig. 1 Total early-stage entrepreneurial activity in India (% adults)

(NECI), India is in the top five economies that actively try to boost the vitality of their entrepreneurial ecosystems by addressing a set of 12 predetermined parameters.

2.1 *Definitions for Science and Technology Entrepreneurship*

In its 2008 report "Entrepreneurship in India," the National Knowledge Commission defined entrepreneurship as:

> Entrepreneurship is the professional application of knowledge, skills and competencies and/or of monetizing a new idea, by an individual or a set of people by launching an enterprise de novo or diversifying from an existing one (distinct from seeking self- employment as in a profession or trade), thus, to pursue growth while generating wealth, employment and social good.

There were also a wide range of definitions of technological entrepreneurship among the 93 articles considered (Table 1).

After Schumpeter [21] gave the term "scientific and technical enterprise" its first meaningful definition, it grabbed the attention of numerous scientists and economists, who proceeded to investigate its logical underpinnings in numerous academic works. The process entails analysing how different business situations call for different sets of knowledge and mental tools [22, 23]. Science and business technology are interpreted in literature as the ability to introduce novel concepts, to emerge with novel technologies, or to expand upon current sales channels to create value for the economy via novel business models [24].

Table 1 List of defining factors of technological entrepreneurship (*Source* Global Entrepreneurship monitor report)

Sl. No	Defining factor	References
1	Organization, management, and risk bearing of a technology-based business	Nicholas and Armstrong [18]
2	Solutions in search of problems	Venkataraman and Sarasvathy [19]
3	Establishment of a new technology venture	Jones-Evans (1995)
4	Ways in which entrepreneurs draw on resources and structures to exploit emerging technology opportunities	Liu et al. [20]
5	Joint efforts to interpret ambiguous data, joint understanding to sustain technology efforts, and persistent, coordinated endeavour to accomplish technological change	Jelinek (1996)
6	An agency that is distributed across different kinds of actors, each of which becomes involved with a technology and, in the process, generates inputs that result in the transformation of an emerging technological path	Garud and Karnøe (2003)

3 Methodology

Research was conducted using both quantitative and qualitative strategies with the goal of defining science and technology entrepreneurship.

To conduct the 'co-authorship and co-occurrence reviews,' we have made use of Waltman et al. [25] VOSviewer programme. When doing a bibliometric analysis, this tool follows the procedures outlined by [26]. The software generates webs of connections, thinking about which journals and terms are most often referenced. When originally started searching for these terms in SCOPUS, got a list of 14,000 results. Through refining these resulting tallys, we were able to arrive at a final tally of 4473 items for our initial tally. The final number of items for the "co-authorship and co-occurrence reviews" was 921 after we removed the duplicates.

The major elements of the current concept of Science and Technology Entrepreneurship have been recognised with the help of the co-occurrence analysis. Science and technology entrepreneurship has been defined in large part due to the efforts of the authors who were afterwards identified. The definition offered by Bailetti [27] is the one most commonly used across all academic works. Consequently, we split our study of defining Science and Technology Entrepreneurship into two time periods: before and after 2012. Seventy-five scholarly works were published on the topic of technological entrepreneurship before 2012 (Table 2).

Table 2 List of number of journal articles, keywords in the title, theme, and period (*Source* Global Entrepreneurship monitor report)

Sl. No.	Themes	Number of articles during the period						Total
		1970–1979	198019–89	1990–1999	2000–2009	2010–2019	Newly identifie d 2010–19	1970–2019
1	Definition of technology entrepreneurship						6	6
2	External factors affecting the existence and survival of technology based firms	0	2	7	23	10	3	45
3	Technology entrepreneurship affecting the regional socio-economic development	0	3	3	7	1	2	16
4	Technology firms characteristics and strategies (3 + 4)	1	1	8	10	2	5	27
5	Interrelation among technology process and growth of technology based firms	0	2	3	1	0	3	9
6	Overview of technology entrepreneurship	1	0	0	2	1	2	6
7	Interrelation of science with technology entrepreneurship						1	1
8	Knowledge intensive entrepreneurship				1		14	15
	Total						36	125

4 Discussion and Findings

The introduction of the 'start-up India initiative' by Prime Minister Shri Narendra Modi on January 16, 2016 has stoked entrepreneurialism across the country. Department for Promotion of Industry and Internal Trade (DPIIT), Union Ministry of Commerce and Industry, has been ranking and mapping state-level start- up activity through their start-up ranking methodology. The institutional assistance, incubation support, seed funding and venture funding support, awareness and outreach initiatives made by the states are all considered within the framework. Both institutional and incubator support measures have been accorded the highest priority. Further, the division has unveiled a platform called "Start-up India Hub," which has data on mentors, support institutions, start-ups, etc., organised by state (Fig. 2).

In order to define science and technology entrepreneurship, these research publications were organised into eight distinct categories. In order to define Science and Technology Entrepreneurship, we collated all the elements proposed by the literature and quickly scanned for any overlap in ideas. Similar considerations led to the identification of nine crucial elements that should be incorporated within the proposed definition. We conducted a study of 35 professionals in the science and technology fields and gathered their replies through questionnaires and in-depth interviews.

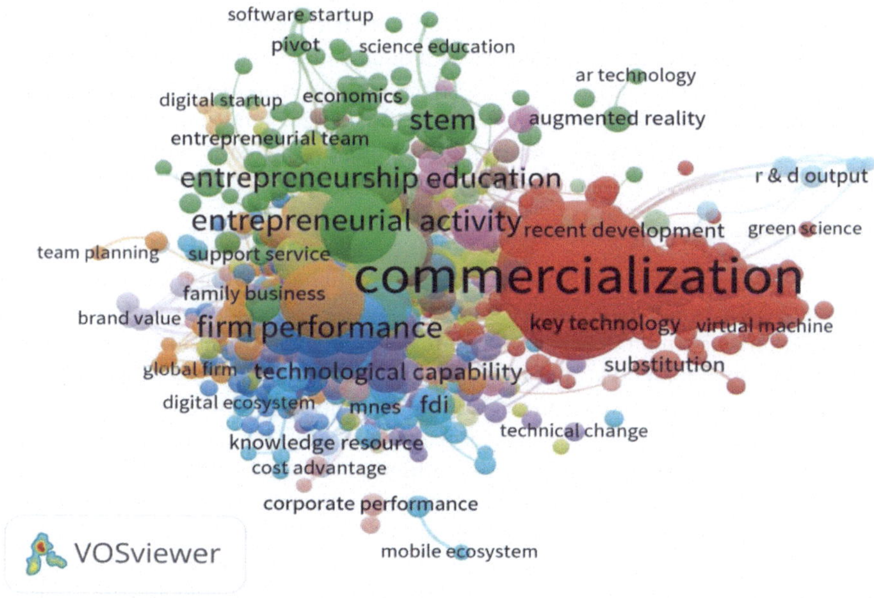

Fig. 2 Co-occurrence analysis of the technology entrepreneurship related terms

4.1 Entrepreneurial Journey of India

India has made great efforts to maintain its position and to take a leap further by mobilising science and technology policies in the country. Because of this, India's economy is now the 52nd largest in all of Asia and the Pacific in 2019. With the world's largest economic growth, India has jumped from the 81st position in 2015 to the 29th position in 2018, despite the fact that it is still a developing nation with one of the highest populations. As a result of its diversification into academic research and development and its contributions to scientific research and development, India continues to rank second in innovation quality among middle-income nations. It consistently ranks near the top on a variety of key metrics, including increase in productivity and exports of ICT-related services. The research and development (R&D) budgets of Indian corporations ranked fifteenth worldwide in 2019. It is also represented in the Global Innovation Index's list of the top 100 scientific and technology clusters in the world, which includes cities like Bengaluru, Mumbai, and New Delhi.

4.2 Incubation Mechanism

The approach to entrepreneurship ecosystem is unique to industrial districts, suburbs and areas with high concentration of tech clusters because the stakeholders in such areas look at entrepreneurs and startups as unique entities with diverse abilities and reserves and with the ability to cope in social and business context. Recent publications on the business environment (Feld 2012) [28, 29] have indicated successful business success and its impact on technology entrepreneurship without providing clear reason and proof of its causes and effects.

Gujarat Entrepreneurship Ecosystem

Gujarat is such an ecosystem where state government had committed to support 500 start-ups at seed stage in a manner of giving them direct grant at the early stage. Generally, this missing link is neither public nor private sector in other state start-up ecosystems. Gujarat had systematic efforts at its pre incubation layer through student start-up interventions across various technology and non-technology universities. Ecosystem actors from non govt. sector has played crucial role in shaping Gujarat ecosystem, particularly in mentoring, early funding, incubation perspective.

More than 200 crore was allocated across various departments of Govt. of Gujarat for supporting start- up/entrepreneurship in last Gujarat Budget. Various schemes have been initiated by the state government for channelizing these financial resources. The schemes are (a) scheme for availing assistance to start-ups/innovations under Gujarat Industrial Policy 2015 recognizing nodal institutions in the state for extending the start-up support, (b) Student Start-up and Innovation Policy, governed by Gujarat Knowledge Society, Directorate of Technical Education, c. Electronics & IT/ITeS Start-up 2016–2021 (Fig. 3).

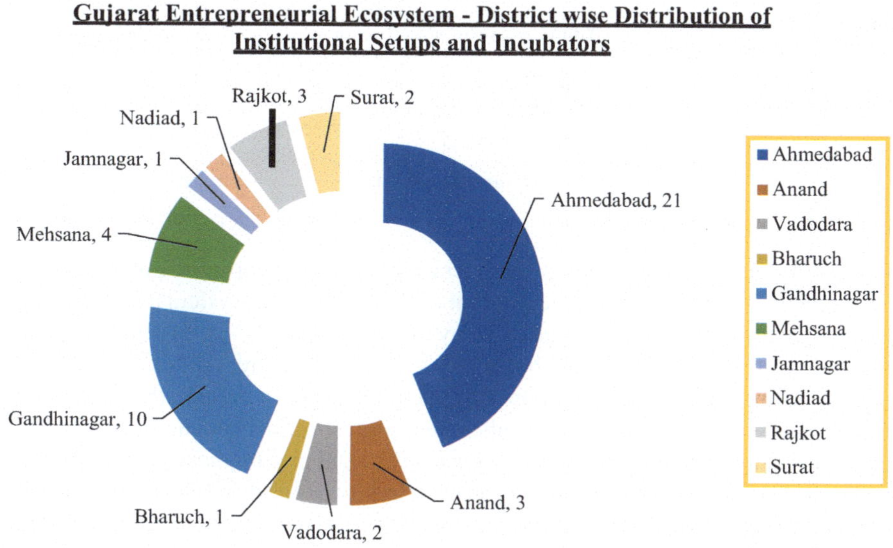

Fig. 3 Gujarat entrepreneurial ecosystem—district wise distribution of institutional setups and incubators

The supply side of the technology and innovation value chain of Gujarat ecosystem is deeply rooted around research, project, and incubation activities across universities in Gujarat. The incubation support structure can be broadly mapped by data Tables 3 and 4.

The term entrepreneurial ecosystem refers to a complex system of highly interconnected activities. According to Ven [30], a lack of understanding of the contribution and engagement of many other ecosystem stakeholders has resulted from the fact that

Table 3 List of different schemes (*Source* Global Entrepreneurship monitor report)

Sl. No.	Scheme/Department supported by	Number of incubation centres
1	Startup Scheme, Industries, Commissionerate, GoG	33
2	Student Startup & Innovation Policy, Directorate of Technical Education, GoG-University Incubators	19
3	DieTY/MieTY	2
4	MHRD	4
5	Department of Biotechnology	2
6	MoMSME	6
7	Science and Technology Parks	4
8	Atal Incubation Centres	4
9	STEP—DST	2

Table 4 List of additional institutional setups (*Source* Global Entrepreneurship monitor report)

Additional institutional setups	Numbers
Fab labs	33
IP facilitation centres	4
Pre-incubation facilities	57
Regional innovation centres	4

the majority of ecosystem study researchers accounted for a specific perspective on entrepreneurs' role. Therefore, the concept of the business ecosystem as a collection of trust elements that are integrated in a way that empowers the entrepreneur appears to be the most pertinent. Several characteristics make the entrepreneurial ecosystem an ideal setting for a variety of supporting activities. These include the facilitation of actual and projected information, the keen interest of investment agencies, skill development and training institutes, governance, legal and financial resources, and the motivation to generate originality and creativity. Recent publications, particularly those with theoretical contributions, provide a practical framework that identifies factors deemed crucial to the success of a start-up ecosystem. Technological business technologies (TBTs) are one of the most essential elements of a modern startup ecosystem [31, 32]. One of the most important interventions for encouraging and inspiring young entrepreneurs and economic development is new venture capital [33] (Innocenti and Zampi 2019). TBI has prioritised the expansion of new domestic companies as a result of the awareness of technology transfer and the enabling of product packaging. By creating entrepreneurial support environments, TBIs enable startups to increase their survival rates through technology adoption [34].

Gujarat is investing a lot in building a strong ecosystem for nurturing innovation and startups to drive economic growth and boost job creation. The state has an ambitious plan to establish around 100 incubators and to develop almost 200,000 m^2 of Incubation Space and support at least 2000 start-ups in the coming years. Over the past two years, Gujarat has provided a financial assistance of INR 220 million (IJSD 3 million) to 175 start-ups. The government has set a target of facilitating 2000 start-ups by 2021. The government has set a target of facilitating 2000 start-ups by 2021. Apart from the 175 government-funded start-ups there are 26 start-ups in the state that are being funded by High Net-worth Individuals (HNIs). Currently there are about 33 incubators in Gujarat against a target of 50 that should be up and running by 2021.

4.3 *Science and Technology Entrepreneurship—Evidence from Global Studies*

One of the main elements of the Singapore development model, and of global STI policies more generally, is the importance of providing assistance to STI-driven enterprises that have viable and adaptable business models. Modern economies, in order to achieve their goals of development and economic progress, place a greater emphasis on entrepreneurship, which in turn leads to the formation of private industry. Countries like India, whose economy is still in its early stages of development, prioritise these characteristics:

(A) A nearby science and research facility is essential for the rapid advancement of new technologies. A mechanism for facilitating market research and evaluating the success of potential research must be established by the Government of India for this to occur.
(B) Putting the emphasis on quick returns on investments made over the short term: It is important to invest in efforts and technology initiatives that can solve the major problem rapidly and in a sustainable way. Motivating academics and teachers to take up careers in emergency response is essential.
(C) Integration of National Research Institutes and Application of All Viable Technologies in the Field of Advanced Scientific Applications: Dynamic Platforms Should Be Provided for Such Interactions This will help small enterprises embrace cutting-edge research and benchmarking global business methods, both of which have lately been integrated into global platforms thanks to increased collaboration and engagement [35].

The scientific community in Europe and other affluent countries is given special attention. Examples include Europe's ambitious Entrepreneurship 2020 Action Plan, which aims to improve business education and training, the business environment, and business models. India updates its proposed business policy since it values entrepreneurship (Fig. 4).

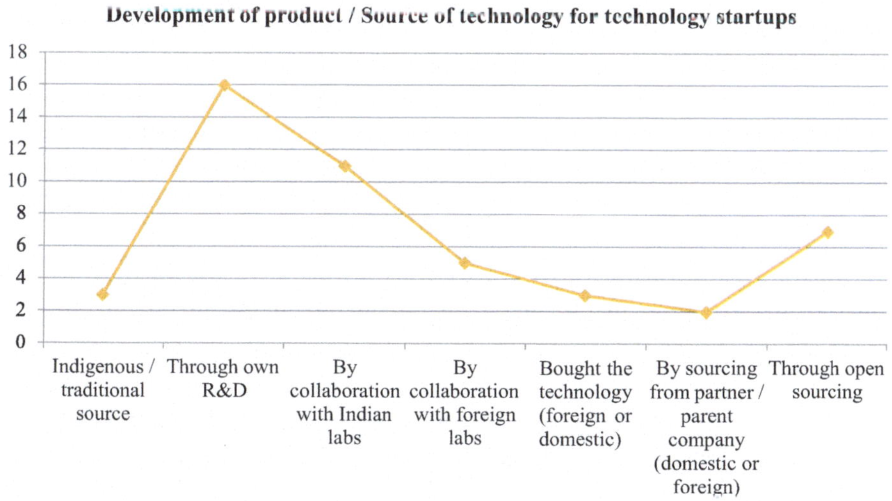

Fig. 4 Product development at various stages

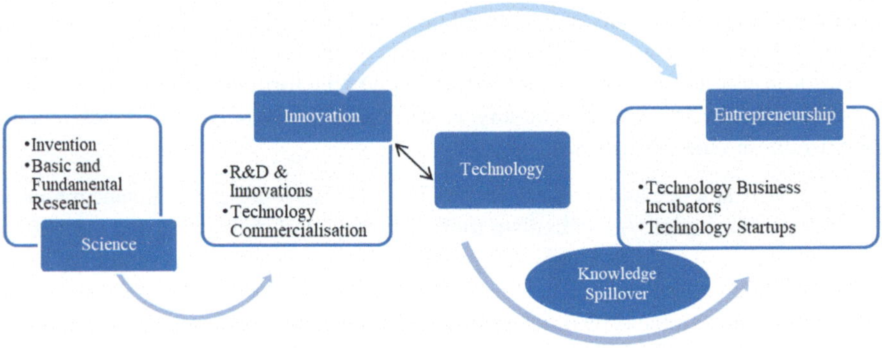

Fig. 5 Entrepreneurship in science and technology: proposed conceptual model

According to the OECD's 2016 report on the outlook for science, technology, and innovation in business, a number of different macroeconomic factors contribute to the development of these fields. Science knowledge, R&D spending, academic research institutions, and research quality are all factors considered in the 2018 edition of the Global Competitiveness Report, which serves as one of the report's four pillars for detecting and mapping technological innovation dynamics. Before, the Global Competitiveness Report divided innovation and technology into two distinct categories when evaluating a country's ability to compete. However, variables like scientific knowledge, expenditure on R&D, research institutions, and quality have converged to form a single pillar of innovation capability as a result of the increasing prevalence of S&T in business R&D and innovation activity (as overall output) (Fig. 5).

There are many different metrics that can be used to measure innovation because it is such a general term. The quantity and quality of an organization's inventive output (the innovations it creates) are both important metrics to consider. Research & development expenditures, patents, and the release of new products or technology are all examples of popular empirical indicators used to gauge scale. Patent citations and the importance of the innovations themselves are indications of their quality. The contributions to economic value through innovation are also measured by how quickly and widely an innovation is adopted. The suggested model is based on a primary survey, much interdisciplinary research, and the indicators gleaned through a review of the relevant literature and discussions with experts in the field.

For decades, science, technology, and innovation have evolved in tandem. Entrepreneurial input and the surrounding support system, which includes institutionalisation, have a significant impact on the development of new technological innovations. It is necessary to develop indicators of mobility between innovators and other organisations, between sectors that produce and use research, and within companies. We have proposed to map the market incentives that motivate technological initiatives such as research and development by companies, which leads to market advantage. This activity also results in the transfer of knowledge, which promotes economic growth and increases economic competitiveness. It essentially describes the role that technology entrepreneurs play by developing new concepts, obtaining patents, and entering markets.

The establishment of an entrepreneurial ecosystem in a given economy relies heavily on high-tech start- ups. As technology transfer and commercialization of technology increase in a nation, there is a significant tendency for the number of high-tech start-ups to rise. TBIs are essential for maintaining the transfer and commercialization of technology from research organisations to industries. In addition, TBIs offer numerous services that assist new entrepreneurs in launching a technology-based venture, start-up, or company. Thus, TBIs are a crucial factor in the expansion of technology-based start-ups. While, after a period of growth, a start-up became a part of industries that directly encourage entrepreneurial endeavours. Consequently, entrepreneurial activities are also useful for increasing economic development and growth, which may be necessary for a rise in the number of new businesses. Consequently, there is a significant relationship between start-ups, technology business incubators, and the entrepreneurial ecosystem in both developed and developing economies.

5 Conclusion

Policy suggestions related to TBIs for mapping their performance and through this process making them the strongest pillar of the entrepreneurial ecosystem. We have observed that TBIs have played a crucial role in enhancing the operational efficiency of start-ups. The importance of a robust framework for mapping the operational support plan for technology start-ups is highlighted. This will serve as an input parameter for monitoring TBI performance.

From the study, it can be inferred that initial seed funding support propels start-ups towards achieving sustainability. This type of support, if channeled by TBIs, strengthens the commitment and responsibility of start-ups to contribute financially. The start-up's submission of a bootstrapping plan within the first six months of the incubation process will aid in determining its commitment and seriousness, as well as its role in ensuring sustainability. This will serve as a process parameter for monitoring TBI performance.

The capacity building interventions by the TBIs in the functional areas have emerged as a very important process parameter for fostering incubate growth. While conducting narrative discussions with the management of TBIs, we discover that the majority of TBIs address innovation and the design-thinking approach. Not enough emphasis was placed on the functional aspects of the start-up process. TBIs are required to share an annual plan of activities for building the start-up's capacity.

Technology business incubators are required to promote exposure to the customer-funded business model [36]. Start-ups should receive training and practical experience to help them comprehend customer-funded models and reduce their reliance on loans and grants. This will function as a process parameter. Another important factor is the management of start-up failures. According to the inputs of LPAC members and the conducted study, the failure of new ventures was considered a performance metric for determining which TBIs should be discontinued. Rather, the management of start-up failure must be utilised as a process parameter. Consequently, we incorporated a variable into the process of pivoting by start-ups. Important sub-parameters for this process include a timely pivot and adequate mentoring support for its implementation.

The output parameter should specify which start-up risks the TBIs have addressed or mitigated. According to the input of LPAC members and the findings of the study, risks have been categorised according to the functions and characteristics of start-ups as follows: operational risk, financial risk, legal risk, technological risk, managerial risk, and market risk. Research and Development conducted within a company in an effort to foster innovation must be identified as a key innovation metric and considered for policy support. This is consistent with Romer's [37] endogenous growth model, for which he was awarded the 2018 Nobel Prize. The technology commercialization process and technology adaptation should be viewed as policy interventions and managed by government-supported institutions.

References

1. Dalrymple, D.: Scientific knowledge as a global public good: Contributions to innovation and the economy. In: The Role of Scientific and Technical Data and Information in the Public Domain: Proceedings of a Symposium, pp. 35–49. National Academic Press, Washington (2003, September)
2. Sorenson, O., Fleming, L.: Science and the diffusion of knowledge. Res. Policy **33**(10), 1615–1634 (2004)
3. Abernathy, W.J., Clark, K.B.: Innovation: mapping the winds of creative destruction. Res. Policy **14**(1), 3–22 (1985)
4. Gittelman, M., Kogut, B.: Does good science lead to valuable knowledge? Biotechnology firms and the evolutionary logic of citation patterns. Manage. Sci. **49**(4), 366–382 (2003)
5. Caraca, J., Lundvall, B.Å., Mendonça, S.: The changing role of science in the innovation process: From Queen to Cinderella? Technol. Forecast. Soc. Chang. **76**(6), 861–867 (2009)
6. Curran, C.S., Bröring, S., Leker, J.: Anticipating converging industries using publicly available data. Technol. Forecast. Soc. Chang. **77**(3), 385–395 (2010)
7. Hacklin, F., Wallin, M.W.: Convergence and interdisciplinarity in innovation management: a review, critique, and future directions. Serv. Ind. J. 33(7–8), S774–S788 (2013)
8. Brooks, F.P: The mythical man-month: Essays on software engineering (Anniversary ed.). Addison-Wesley (1994)
9. Knight, F.H.: Cost of production and price over long and short periods. J. Polit. Econ. **29**(4), 304–335 (1921)
10. Schumpeter, J.A.: Capitalism, socialism, and democracy (3rd ed.). Harper & Brothers (1950)
11. Berna, J.J.: Patterns of Entrepreneurship in south India. Econ. Dev. Cult. Change **7**(3, Part 1), 343–362 (1959)
12. Lamb, H.B.: The Indian business communities and the evolution of an industrialist class. Pac. Aff. **28**(2), 101–116 (1955)
13. Medhora, P.B.: Entrepreneurship in India. Polit. Sci. Q. **80**(4), 558–580 (1965)
14. Rungta, R.S.: The rise of business corporations in India, 1851–1900 (No. 8). CUP Archive (1970)

15. Tripathi, D.: Indian entrepreneurship in historical perspective: a re-interpretation. Econ. Polit. Week., M59–M66 (1971)
16. Watanabe, S.: Entrepreneurship in small enterprises in Japanese manufacturing. Int. Lab. Rev. **102**, 531 (1970)
17. Kilby, P.: (ed.): Entrepreneurship and economic development. Free Press (1971)
18. Nicholas, S., Armstrong, J.: Management knowledge and the new millennium. Oxford University Press (2003)
19. Venkataraman, S., Sarasvathy, S.D.: Strategy and entrepreneurship: Outlines of an untold story. In D.O. Faulkner, M. de Rond (eds.), Cooperative strategy: Economic, business, and organizational issues, pp. 650–668. Oxford University Press (2000)
20. Liu, W., Huang, G., Liu, X., Min, Z., Deng, X.: Detection of phishing webpages based on visual similarity. International World Wide Web Conference, pp. 1060–1061 (2005)
21. Schumpeter, J.A.: The theory of economic development: An inquiry into profits, capital, credit, interest, and the business cycle (R. Opie, Trans.). Harvard University Press. (Original work published in 1911) (1934)
22. Malerba, F.: (ed.): Knowledge intensive entrepreneurship and innovation systems: evidence from Europe, Vol. 48. Routledge (2010)
23. Venkataraman, S.: Regional transformation through technological entrepreneurship. J. Bus. Ventur. **19**(1), 153–167 (2004)
24. Romano, A., Passiante, G., Vecchio, P.D.: The technology-driven entrepreneurship in the knowledge economy. In: Creating Technology-Driven Entrepreneurship, pp. 21–48. Palgrave Macmillan, London (2016)
25. Waltman, L., Calero-Medina, C., Kosten, J., Noyons, E.C., Tijssen, R.J., van Eck, N.J., Wouters, P.: The Leiden ranking 2011/2012: data collection, indicators, and interpretation. J. Am. Soc. Inform. Sci. Technol. **63**(12), 2419–2432 (2012)
26. Cobo, M.J., López-Herrera, A.G., Herrera-Viedma, E., Herrera, F.: Science mapping software tools: review, analysis, and cooperative study among tools. J. Am. Soc. Inform. Sci. Technol. **62**(7), 1382–1402 (2011)
27. Bailetti, T.: Technology entrepreneurship: overview, definition, and distinctive aspects. Technol Innov Manage Revi. **2**(2) (2012)
28. Isenberg, D.: The entrepreneurship ecosystem strategy as a new paradigm for economic policy: principles for cultivating entrepreneurship. Present. Inst. Int. Eur. Affairs **1**(781), 1–13 (2011)
29. Spigel, B.: The relational organization of entrepreneurial ecosystems. Entrep. Theory Pract. **41**(1), 49–72 (2017)
30. Van de Ven, H.: The development of an infrastructure for entrepreneurship. J. Bus. Ventur. **8**(3), 211–230 (1993)
31. Alsos, G.A., Carter, S., Ljunggren, E.: (eds.): The handbook of research on entrepreneurship in agriculture and rural development. Edward Elgar Publishing (2011)
32. Bala Subrahmanya, M.H.: How did Bangalore emerge as a global hub of tech start-ups in India? Entrepreneurial ecosystem—evolution, structure and role. J. Dev. Entrep. **22**(01), 1750006 (2017)
33. Corsi, C., Di Berardino, D.: Assessing the business incubators' performance referring the local development in Italy. Eur. Sci. J. ESJ, **10**(10) (2014)
34. Aerts. K., Matthyssens, P., Vandenbempt, K.: Critical role and screening practices of European business incubators. Technovation **27**, 254–267 (2005). https://doi.org/10.1016/j.technovation.2006.12.002
35. Lerner, J.: The future of public efforts to boost entrepreneurship and venture capital. Small Bus. Econ. **35**(3), 255–264 (2010)
36. Mullins, J.: The counter-conventional mindsets of entrepreneurs. Business Horizons, **60**(5), 597–601 (2015). https://doi.org/10.1016/j.bushor.2015.04.002
37. Romer, P.M.: The problem of development: A conference of the Institute for the study of free enterprise systems. J. Polit. Econ. **98**(5), S71–S102 (1990)
38. Capone, F., Innocenti, N.: Open innovation and network dynamics: An analysis of openness of co-patenting collaborations in Florence, Italy. Comp. Rev.: Int. Bus. J. ahead-of-print (2020) https://doi.org/10.1108/CR-10-2019-0101

Harnessing Strategic Analytics for Start-Up Success: Innovations, Insights, and Impact

S. Hemanth Kumar, S. Mahalakshmi, S. Kiran Maney, and Sandeep Kautish

Abstract Strategic analytics has become crucial for start-up organizations aiming for growth and sustainability. This chapter explores the core elements of strategic analytics, including its definition, scope, historical development, and key theories and models. It highlights the essential analytics' function in locating market opportunities, conducting competitive analyses, gaining deep customer insights. By examining case studies of successful start-ups, this chapter demonstrates the transformative power of data-driven decision-making and offers best practices for leveraging analytics to drive innovation, optimize operations, and build resilience. Entrepreneurs are encouraged to invest in strong analytics capabilities, foster a data-driven culture, ensure ethical data use, and remain agile in the face of technological advancements and market disruptions.

Keyword Strategy · Analytics · Sustainability · Data-driven decision-making · Entrepreneurship · Market disruptions

1 Introduction

1.1 Importance of Strategic Analytics for Start-Ups

In today's data-centric landscape, strategic analytics is now a crucial instrument for start-ups aiming to navigate market complexities and achieve sustainable growth. The ability to harness large datasets and derive actionable insights provides start-ups with a competitive edge, enabling them to continuously innovate, optimize processes, and

S. Hemanth Kumar · S. Mahalakshmi (✉) · S. Kiran Maney
Faculty of Management Studies, CMS Business School, JAIN (Deemed-to-Be) University, Bangalore, India
e-mail: dr.mahalakshmibharath@gmail.com; mahalakshmi.s@cms.ac.in

S. Kautish
AIT CSE, Chandigarh University, Mohali, Punjab, India

S. Kautish et al. (eds.), *Strategy Analytics for Business Resilience Theories and Practices*, Information Systems Engineering and Management 33,
https://doi.org/10.1007/978-3-031-82369-5_8

make well-informed decisions [25]. Through strategic analytics, start-ups can identify market trends, understand customer behaviors, and anticipate future challenges, thereby positioning themselves effectively in the market [85].

Strategic analytics is more than just a decision-making tool; it is a fundamental component of a start-up's strategy. It empowers entrepreneurs to base their strategies on empirical evidence rather than intuition, thus minimizing risks and enhancing the likelihood of success [74]. In addition, start-ups can discover fresh prospects and promote innovation by using contemporary analytical technologies such as big data analytics, machine learning, and artificial intelligence [77]. By leveraging these technologies, start-ups can achieve a deeper understanding of their operational environment and create more resilient and adaptive business models.

2 Foundations of Strategic Analytics

2.1 Definition

In order to support and enlighten strategic business choices, data, statistical analysis, and technology are systematically used in strategic analytics. It encompasses activities such as data collection, processing, and the application of advanced analytical techniques to extract actionable insights from data. The main goal of strategic analytics is to align data-driven insights with business objectives, enabling organizations to optimize their strategies, operations, and performance.

2.2 Scope

The scope of strategic analytics is broad, covering various organizational domains. Data Management entails acquiring, retaining, and organizing substantial amounts of organized and unorganized data from various sources. Descriptive Analytics provides a summary of historical data in order to gain insights into prior occurrences and patterns. Diagnostic Analytics conducts data analysis to ascertain the causal factors of previous events. Predictive Analytics forecasts future trends and outcomes using machine learning algorithms and statistical models [57]. Prescriptive Analytics offers actionable recommendations based on predictive insights to achieve desired outcomes [25]. Data Visualization represents data and insights through visual formats like charts, graphs, and dashboards to facilitate understanding and decision-making [35].

2.3 Historical Context and Evolution

The foundations of strategic analytics have evolved significantly over time, beginning with early statistical methods and advancing through various technological innovations. Early Statistical Methods involve the use of basic statistical techniques in business analysis, dating back to the early twentieth century, focusing on methods like regression analysis and hypothesis testing [37]. The 1960s and 1970s saw the advent of Management Information Systems (MIS), which provided managers with essential data for decision-making as computers became more prevalent [82]. During the 1980s and 1990s, the rise of Business Intelligence (BI) tools and systems enabled efficient data collection, storage, and analysis, primarily focusing on descriptive analytics and reporting [125]. The 1990s also witnessed the emergence of Data Warehousing technology, allowing corporations to merge data from several sources into a unified repository for thorough analysis [59]. The 2000s were defined by the emergence of Big Data and Advanced Analytics, involving the utilization of intricate algorithms, machine learning, and artificial intelligence to examine extensive and varied datasets [14].

2.4 Essential Theories and Models in Strategic Analytics

Strategic analytics encompasses important ideas and models, one of which is Decision Theory. This theory offers a structured approach to making well-informed decisions when faced with uncertain situations. It incorporates notions like expected utility and risk assessment, as defined Systems Theory is a perspective that regards organizations as intricate systems made up of interconnected components, emphasizing the significance of comprehending interactions and interdependencies within the system for efficient decision-making [13]. The Balanced Scorecard (BSC) is a strategic planning and management method developed by Kaplan and Norton, combining financial and non-financial performance measurements from four perspectives—financial, customer, internal processes, and learning and growth—to synchronize company actions with vision and strategy [65]. Value Chain Analysis, proposed by Porter, identifies the primary and support activities that create value for an organization, helping to understand how data analytics can optimize each activity to enhance overall performance [97]. Predictive Modeling involves forecasting future occurrences using statistical methods and machine learning algorithms based on prior data, with commonly used models including logistic regression, decision trees, neural networks, and linear regression [61].

3 The Role of Strategic Analytics in Start-Up Ecosystems

3.1 Market Opportunity Identification

Identifying market possibilities is essential for the success and expansion of start-ups. Strategic analytics offers the required tools and methodology to discover these chances by evaluating industry trends, customer behavior, and competition dynamics. Market analysis encompasses various essential methodologies. Trend Analysis, which includes analyzing historical data and current market trends, helps start-ups identify emerging opportunities through methods such as time series analysis and market basket analysis, aiding in understanding consumer preferences and predicting future trends [1]. Sentiment Analysis, which involves forecasting future occurrences, utilizes statistical approaches and machine learning algorithms based on past data, with commonly used models including logistic regression, decision trees, neural networks, and linear regression [93]. Gap Analysis enables start-ups to detect market gaps by comparing market demand with existing products and services, analyzing competitor offers to discover unmet requirements and opportunities for innovation [69].

3.2 Methods for Doing Competitive Analysis

Competitive analysis techniques are crucial for start-ups to comprehend their market position and strategize efficiently. SWOT Analysis assists start-ups in evaluating their competitive position by identifying internal strengths and weaknesses, as well as external opportunities and threats [47]. Porter's Five Forces model considers competitive rivalry, the threat of substitutes, the bargaining power of suppliers and consumers, and the threat of new entrants as key elements in understanding the competitive dynamics within an industry. This model helps start-ups comprehend the intricacies of the business environment and identify strategic opportunities [96]. Benchmarking, which involves contrasting a start-up's performance metrics with those of industry leaders, helps pinpoint areas for improvement and best practices. Frequently benchmarked key performance indicators (KPIs) include market share, customer satisfaction, and operational efficiency [12].

3.3 Methods for Obtaining Customer Insights and Dividing Customers into Segments

Customer insights and segmentation techniques are crucial for firms seeking to improve customer experience and optimize marketing strategies. Customer Journey Mapping is the systematic monitoring and assessment of the various points of contact that consumers come across, starting from the first awareness stage and continuing through to post-purchase. This process enables the identification of both challenges and potential areas for enhancing the overall client experience [104]. Cluster Analysis helps in identifying discrete market segments and adjusting marketing initiatives to meet the unique requirements of each segment [66]. Predictive Analytics for Customer Behavior utilizes past data to make predictions about future consumer behavior, including the likelihood of purchase, risk of churn, and lifetime value, aiding in the creation of focused marketing campaigns and customized customer experiences [79]. Case studies such as Netflix and Spotify illustrate the application of these techniques. Netflix employs customer insights and predictive analytics to customize content recommendations by examining viewing patterns and segmenting users based on viewing habits, leading to enhanced user engagement and increased subscriber retention [41]. Similarly, Spotify leverages customer data to improve user experience and engagement through personalized music recommendations and user behavior analysis, resulting in increased user satisfaction and retention [129].

3.4 Data-Driven Decision Making for Start-Ups

Building a data-driven culture is essential for start-ups to effectively utilize data analytics for strategic decision-making. This involves integrating data into every decision, promoting data literacy among employees, and ensuring access to high-quality data. Encouraging data literacy means ensuring that every employee understands the value of data and knows how to utilize it wisely, achieved through continual learning opportunities, workshops, and training programs [26]. Promoting data-informed decision-making requires incorporating data analytics into core business operations, setting clear objectives for data use, developing performance indicators, and establishing a feedback loop for continuous improvement based on data insights [99]. Ensuring data accessibility and quality involves investing in a robust data infrastructure, implementing data governance procedures, and ensuring data is accurate, comprehensive, and consistent to support a thriving data-driven culture [86].

3.5 Technologies and Tools for Data Analytics

Gathering and storing data effectively is crucial for start-ups, with contemporary databases and data warehousing solutions like PostgreSQL, MySQL, and Amazon Redshift being essential for handling massive volumes of data [120]. Big data technologies such as Hadoop and Spark facilitate managing large datasets in distributed computing environments, crucial for handling big data [45]. Business intelligence tools like Tableau, Power BI, and QlikView enable data visualization and the creation of interactive dashboards for intuitive data insights exploration [109]. Statistical and analytical software such as R, Python, and SAS provide powerful capabilities for statistical analysis and machine learning, essential for complex data analyses and predictive modeling [126]. ETL tools like Talend and Apache NiFi assist in integrating data from various sources, ensuring it is clean and ready for analysis [134]. Data management platforms such as Snowflake and DataBricks offer comprehensive solutions for managing, integrating, and analyzing data at scale. Successful start-ups leveraging data analytics include Uber, which uses predictive analytics to forecast demand and adjust driver availability, employing machine learning algorithms for dynamic pricing based on real-time supply and demand [15]. Airbnb personalizes user experiences and improves its recommendation system by analyzing user data to offer personalized travel suggestions and optimize search results, enhancing customer satisfaction and engagement [46]. Netflix uses advanced algorithms and machine learning to analyze viewing patterns and recommend content tailored to individual user preferences, significantly contributing to its success in retaining subscribers and expanding its user base [41].

3.6 Novel Methods for Strategic Analytics

Machine learning (ML) and artificial intelligence (AI) are revolutionizing strategic analytics by allowing startups to analyze large volumes of data, discover concealed patterns, and enhance forecast precision. This automation of decision-making processes provides a competitive advantage in fast-changing markets. Machine learning in analytics leverages statistical models and algorithms, allowing computers to perform tasks without explicit instructions through patterns and inference, with methods including reinforcement learning, supervised learning, and unsupervised learning [63]. Supervised learning involves using labeled data to train models for predicting outcomes, employing algorithmic models such as support vector machines, decision trees, and linear regression [89]. Unsupervised learning analyzes unlabeled data to uncover hidden patterns, utilizing techniques like principal component analysis and dimensionality reduction (such as k-means) to identify underlying structures [48]. Reinforcement learning trains models to make a series of decisions by rewarding desired actions, particularly useful in dynamic environments where the model learns optimal actions over time [121]. Artificial intelligence in analytics encompasses technologies that enable machines to mimic human intelligence, including machine learning, natural language processing (NLP), and computer vision, thereby automating complex tasks, enhancing predictive accuracy, and extracting deep insights from unstructured data. NLP extracts valuable information from textual data through analysis, with techniques like sentiment analysis and topic modeling improving understanding of customer sentiments and trends [84]. Computer vision analyzes visual data to extract insights, with applications such as image recognition, object detection, and video analysis used for quality control and surveillance [42]. Predictive analytics employs historical data to forecast future outcomes, using techniques like regression analysis, time series analysis, and machine learning algorithms to make accurate predictions about customer behavior, market trends, and operational performance [114]. Regression analysis identifies relationships between variables and predicts outcomes using techniques such as linear and logistic regression [61]. Time series analysis examines data points over time to identify trends and seasonal patterns using methods like ARIMA and exponential smoothing [57]. Machine learning algorithms apply to complex datasets for prediction using neural networks, gradient boosting, and random forests [40]. Prescriptive analytics extends beyond predicting future events to provide specific recommendations on achieving desired outcomes, employing optimization and simulation methodologies to determine the best course of action [7]. Optimization uses mathematical methods like linear programming and integer programming to find optimal solutions [52], while simulation techniques such as Monte Carlo simulation model the impact of different decisions and uncertainties [107].

4 Real-Time Analytics Enables the Swift and Efficient Processing of Data to Facilitate Agile Decision Making

Real-time analytics refers to the prompt processing and examination of data as it is produced, allowing for swift decision-making and timely insights. This is particularly important for start-ups functioning in dynamic situations where quick response is crucial. Technologies for real-time analytics include stream processing tools like Apache Kafka and Apache Flink, which allow data streams to be processed in real-time for swift analysis and response [70], and in-memory computing solutions such as Redis and Hazelcast, which store data in RAM to significantly speed up data retrieval and analysis [45]. Applications of real-time analytics include enhancing operational efficiency by monitoring and optimizing operations like supply chain management, providing instant insights into inventory levels and demand patterns, and improving customer experience through real-time personalization and immediate response to customer inquiries and issues [25]. For example, Walmart uses real-time analytics to streamline their supply chain processes by analyzing data from point-of-sale systems, social media, and weather forecasts, enabling them to forecast demand and adjust inventory levels instantly to ensure products are available when and where customers need them [49].

5 Strategies for Implementing Innovative Analytics Methods in Start-Up Companies

Disruptive analytics tactics for start-ups, such as growth hacking, entail rapidly experimenting with different marketing channels and product development strategies to determine the most effective approaches for firm expansion. Data insights are crucial in growth hacking, providing the necessary information for optimizing strategies and achieving scalable growth. Data-driven growth strategies include split testing, which compares the effectiveness of different versions of a webpage, feature, or marketing campaign by analyzing user interactions and conversion rates, enabling start-ups to make informed decisions about which variant to implement [68]. Tracking customer interactions involves using analytics to monitor user interactions across multiple touchpoints, mapping the customer journey to identify problems and improve user experience, thereby increasing customer satisfaction and conversion rates [104]. Cohort analysis, which divides users into groups based on common traits or behaviors and tracks their activities over time, helps in understanding user retention and identifying factors that drive sustained engagement.

5.1 Utilizing Large-Scale Data Analysis to Gain a Competitive Edge

Utilizing big data to gain a competitive edge enables start-ups to make informed decisions and discover novel prospects by emphasizing connections, patterns, and trends, especially in relation to human behavior and interactions. Big data's key applications include market analysis, which examines market trends, customer preferences, and competitive dynamics to help start-ups identify emerging opportunities and make strategic decisions based on comprehensive market insights [85]. Personalization, achieved by evaluating consumer data to tailor goods, services, and marketing messages to individual preferences, enhances customer satisfaction and loyalty. Predictive maintenance, through big data analytics applied to sensor data analysis and pattern recognition, anticipates equipment failures in manufacturing and other industries, reducing downtime and maintenance costs [76]. The Internet of Things (IoT) represents an interconnected network of devices that communicate and share data, which can be analyzed to provide valuable insights for optimizing operations, improving customer experiences, and fostering innovation. IoT analytics applications include smart products that gather and communicate data regarding usage and performance, offering real-time monitoring and insights for improvement. Operational efficiency is enhanced by utilizing IoT analytics to improve supply chain and logistics operations, providing immediate insights into inventory levels, transportation routes, and environmental conditions, leading to cost reductions and increased efficiency [130]. Predictive analytics, when integrated with IoT data, accurately predicts future trends and occurrences, such as equipment breakdowns, energy consumption patterns, and client demand swings, enabling proactive measures to address issues and optimize performance. Case studies of successful start-ups leveraging disruptive analytics strategies include Spotify, which uses vast datasets and machine learning to tailor music recommendations to individual users by analyzing listening behavior, thereby improving user engagement and loyalty [129]. Tesla enhances vehicle performance and safety by leveraging IoT and big data analytics, collecting data from sensors to improve features, perform predictive maintenance, and provide over-the-air software updates, ensuring continuous improvement and customer satisfaction [51]. Uber employs real-time analytics to enhance its ride-hailing service efficiency by analyzing data from GPS, ride requests, and traffic patterns to dynamically adjust prices, reduce wait times, and improve service efficiency, providing a superior customer experience [15].

5.2 Enhancing Resilience with Analytics

Improving resilience through the use of analytics entails implementing efficient risk management strategies, where strategic analytics aids in identifying, evaluating, and reducing risks, enabling start-ups to anticipate potential risks, develop mitigation strategies, and navigate uncertainties. Data-driven risk identification uses analytics to analyze historical data, market trends, and internal operations, with techniques like data mining and anomaly detection uncovering potential threats [101]. Risk assessment models, such as Monte Carlo simulations and Value at Risk (VaR), provide a probabilistic approach to evaluating the potential impact of risks, helping estimate the likelihood and severity of different scenarios. Predictive analytics anticipates and estimates possible hazards and their consequences, allowing for preemptive actions to minimize their impact, for example, recognizing patterns that indicate financial difficulties to enable proactive measures [114]. Operational risk management, through the analysis of operational data, pinpoints inefficiencies and weaknesses in processes, and implementing corrective measures based on these insights helps reduce operational risks and improve overall efficiency [54]. Scenario planning and simulation equip start-ups with tools to navigate uncertainty and unforeseen circumstances, investigating various scenarios to evaluate potential consequences and formulate strategies. Strategic foresight involves creating detailed narratives about possible future states based on current trends and uncertainties, helping start-ups identify challenges and opportunities and develop flexible strategies [113]. Scenario development uses a blend of qualitative and quantitative data, with analytical tools like PEST and SWOT analyses [18]. Simulation techniques, such as Monte Carlo simulation, employ stochastic sampling and statistical modeling to approximate the likelihood of various outcomes, valuable for evaluating financial vulnerabilities and operational ambiguities [107]. Agent-based modeling simulates the activities and interactions of autonomous agents to assess their influence on the system, useful for understanding intricate systems and forecasting the consequences of various tactics [83].

5.3 Adaptive Strategies for Uncertain Environments

Start-ups must employ adaptive methods in order to effectively navigate uncertain situations. These techniques involve consistently monitoring the company environment, employing flexible decision-making processes, and swiftly adjusting course in reaction to new information. Continuous monitoring and feedback loops are crucial, with real-time analytics allowing start-ups to continuously monitor key performance indicators (KPIs) and respond swiftly to market or operational changes, helping identify emerging risks and opportunities [85]. Establishing feedback loops involves collecting data from various sources, analyzing it, and using the insights to inform decision-making, ensuring strategies remain pertinent and efficient despite shifting circumstances. Flexible decision-making processes are supported by agile methodologies, which enable start-ups to stay flexible and responsive through practices like iterative development and cross-functional teams, facilitating rapid adaptation to new information and changing priorities [105]. Decision-making frameworks such as the OODA Loop (Observe, Orient, Decide, Act) provide structured approaches for making quick, informed decisions in dynamic environments, helping maintain situational awareness and adapt strategies as needed. Case studies demonstrate how resilience can be built through analytics, such as Amazon employing data-driven risk management techniques to enhance supply chain efficiency by predicting future demand and managing inventory levels, thus minimizing stockouts and overstocking [25]. Google uses scenario planning and simulation techniques to predict and prepare for shifts in the technology industry, formulating flexible strategies to ensure long-term competitiveness [58]. Netflix maintains resilience by implementing adaptive techniques and continuous monitoring, adjusting its content offerings based on real-time analysis of viewer data, which allows it to respond promptly to changing viewer tastes and maintain flexibility in adapting to market developments [41].

6 Product Development Strategic Analytics

Strategic analytics for product development entails utilizing data to design consumer-centric goods that specifically respond to distinct customer requests and preferences. This approach ensures that products resonate with the target audience and achieve higher market acceptance. Understanding customer needs is paramount, and analyzing customer feedback and survey data helps comprehend customer preferences, pain points, and desired features, with techniques such as text mining and sentiment analysis providing significant insights [81]. Behavioral data, including website browsing, purchase history, and product usage, offers valuable insights into how customers interact with products, guiding the design of features that enhance user experience [34]. Customer journey mapping, which involves mapping the entire customer experience from awareness to post-purchase, identifies critical touchpoints and areas for improvement, ensuring a seamless and satisfying customer experience

[104]. In the MVP (Minimum Viable Product) process, strategic analytics ensures that the MVP effectively tests key hypotheses and provides actionable insights for further development. Identifying core features is crucial, and analytics helps prioritize features based on customer feedback, market demand, and competitive analysis, with techniques such as conjoint analysis and A/B testing determining the most valued features [44]. Usage data analysis from early product usage helps identify frequently used features, optimizing the MVP and prioritizing those that deliver the highest value to users [106]. Validating product assumptions involves testing key hypotheses about the product and its market, with analytics assessing the accuracy of these assumptions using indicators like user engagement, conversion rates, and customer happiness [8]. Implementing customer feedback loops is crucial for testing assumptions and making adjustments by gathering and analyzing feedback from MVP customers, ensuring that the product evolves in line with customer needs and expectations [23].

7 Iterative Development and Feedback Loops

Iterative development refers to a process where a project is broken down into smaller, manageable parts that are developed and tested in cycles. Feedback loops are an essential component of this process, as they allow for continuous evaluation and improvement based on input from stakeholders and users. The process of iterative development and feedback loops is crucial for continuously improving a product by incorporating user feedback and performance data, thereby guaranteeing that it meets the needs of customers.

Implementing iterative development involves agile methodologies like sprints and retrospectives, which facilitate ongoing improvements, with analytics tracking progress, measuring outcomes, and identifying areas for enhancement in each iteration [105]. Performance metrics such as key performance indicators (KPIs), user retention, feature adoption, and Net Promoter Score (NPS) provide quantitative measures of product performance [103]. Creating effective feedback loops involves establishing channels for continuous user feedback, such as in-app surveys and user testing sessions, ensuring a steady stream of insights to inform product improvements [71]. A data-driven approach to iterations, informed by data collected from previous cycles, ensures that product development is guided by real user experiences and market conditions, leading to more effective and customer-centric solutions [95]. Successful start-ups leveraging strategic analytics in product development include Dropbox, which used an MVP approach to validate its product concept by creating a simple video demonstrating the product's functionality, gathering significant interest and feedback before heavy development investment, with analytics playing a crucial role in understanding user behavior and enhancing the product [55]. Slack employed iterative development and robust feedback loops to continuously improve its communication platform by analyzing user data and feedback, identifying key features that drove user engagement, and iteratively enhancing its product, leading to rapid adoption and growth [10]. Spotify leverages customer-centric product design and iterative development to enhance its music streaming service by analyzing user data and conducting A/B tests, tailoring its features to meet user preferences and ensuring high levels of user satisfaction and engagement [129].

8 Analytics for Marketing and Sales in Start-Up Companies

Utilizing digital marketing analytics is crucial for start-ups to optimize strategies, gain insights into customer behavior, and maximize return on investment (ROI) in marketing and sales. Digital marketing analytics encompasses the quantification, administration, and examination of data derived from online marketing endeavors. Key metrics and tools in this area include website analytics, where tools like Google Analytics provide valuable information on website traffic, user engagement, and conversion rates, with metrics such as page views, bounce rate, and average session length offering significant insights into user behavior and highlighting areas for improvement [20]. Social media analytics involves using tools provided by social media platforms to monitor and measure various engagement indicators, including likes, shares, comments, and follower growth, with tools like Hootsuite and Sprout Social aiding in analyzing the effectiveness of social media campaigns [94]. Email marketing analytics encompasses the collection and analysis of data related to open rates, click-through rates, and conversion rates from platforms like Mailchimp and HubSpot, helping to optimize email campaigns for better engagement and effectiveness [16]. By utilizing these analytics tools and metrics, start-ups can gain a comprehensive understanding of their marketing efforts, make data-driven decisions, and achieve higher levels of customer engagement and sales performance.

9 Enhancing the Efficiency of Marketing Campaigns

Optimizing marketing campaigns for start-ups entails utilizing digital marketing analytics to assess and improve different facets of online marketing endeavors, guaranteeing increased engagement and improved return on investment (ROI). A/B testing evaluates two variations of a marketing asset to determine which one yields superior results, allowing start-ups to test different headlines, images, and call-to-actions to improve engagement and conversion rates [68]. Attribution modeling helps understand which marketing channels lead to conversions, with models like first-touch, last-touch, and multi-touch attribution providing insights into the customer journey and aiding in effective allocation of marketing funds [80]. Analysis of sales performance, involving the assessment of sales methods and identification of areas for improvement, is critical for driving revenue growth. Key performance indicators (KPIs) such as sales conversion rate, average deal size, and sales cycle length are used to measure sales effectiveness, optimize processes, and accelerate revenue generation [62]. Sales process optimization includes pipeline analysis to identify stages where deals are lost and bottlenecks in the sales process, and sales forecasting to predict future sales and plan inventory, set targets, and make informed business decisions [17, 117]. Customer journey mapping involves creating visual representations of the different phases customers go through, from awareness to post-purchase, to identify critical touchpoints and areas for improvement [78]. Collecting data from various touchpoints such as website visits, social media interactions, and customer service requests provides a holistic view of the customer journey, while identifying touchpoints and analyzing customer behavior helps understand movement through the journey and pinpoint preferences and pain points [21, 31]. Optimizing the customer journey through personalization and addressing pain points enhances customer satisfaction and loyalty, significantly improving conversion rates and retention [104]. Successful start-ups like Airbnb, HubSpot, and Amazon leverage these strategies effectively. Airbnb uses digital marketing analytics to optimize its online marketing initiatives, enhancing user acquisition and retention by analyzing website traffic, user behavior, and conversion rates [46]. HubSpot employs sales performance analysis to continuously improve its sales processes, boosting revenue by monitoring KPIs like conversion rates, transaction size, and sales cycle duration. Amazon uses customer journey mapping to improve the shopping experience, simplifying the purchase process, enhancing customer satisfaction, and personalizing recommendations by analyzing customer interactions and touchpoints [25].

10 Analysis of Financial Data and Development of Strategies for Obtaining Funding

Financial analytics and funding strategies are essential for the financial well-being and strategic planning of start-ups. They enable informed investment decisions, effective cash flow management, and accurate prediction of future revenue through precise financial forecasts. Budgeting also helps with efficient resource allocation and the establishment of financial objectives. Techniques for financial forecasting include time series analysis, which uses historical financial data to forecast future performance with prevalent techniques like ARIMA (AutoRegressive Integrated Moving Average) and exponential smoothing [57], scenario analysis that develops various financial scenarios based on different future assumptions to help start-ups prepare for potential uncertainties and make strategic decisions under different conditions [113], and financial modeling that uses mathematical models to represent the financial performance of a business, with tools like Excel and financial modeling software aiding in building models to project revenues, expenses, and cash flows [5]. Budgeting strategies include zero-based budgeting (ZBB), a budgeting approach that necessitates justifying all expenses for each new period from a baseline of zero, ensuring resources are distributed according to present requirements rather than past expenditures [100], and activity-based budgeting (ABB), which focuses on budgeting for the activities that generate expenditures, establishing a connection between budgeting and specific business activities and processes, thereby enhancing cost control and efficiency [64].

11 Investor Relations and Fundraising Analytics

Effective investor relations and fundraising strategies are crucial for securing capital and building trust with investors, with analytics playing a vital role in communicating financial performance, optimizing fundraising efforts, and understanding investor behavior. Analyzing investor behavior involves investor segmentation, where investors are segmented based on preferences, investment behaviors, and risk profiles, using demographic, geographic, and psychographic data to target the right investors for fundraising campaigns [75]. Investor sentiment analysis of communications such as emails, social media posts, and meeting transcripts helps understand investor sentiments and concerns, enabling tailored communications that address investor needs [90]. Optimizing fundraising efforts through campaign analytics involves monitoring and evaluating fundraising campaigns with metrics like conversion rates, average donation size, and donor retention rates to provide insights into the efficacy of fundraising strategies [111]. Predictive analytics for fundraising identifies potential investors more likely to invest based on their past behavior and engagement, allowing efforts to be concentrated on high-probability prospects . Communicating financial performance through regular and transparent reporting builds trust with investors, with analytics tools automating the creation of detailed financial reports, dashboards, and visualizations to effectively communicate key performance indicators (KPIs) [32]. Interactive dashboards created with tools like Tableau and Power BI allow investors to dynamically explore financial data, providing real-time insights into financial performance [109].

12 Financial Health Monitoring

Monitoring the financial health of a start-up is crucial for ensuring sustainability and growth, involving regular assessment of financial performance, identification of potential issues, and informed decision-making to maintain stability. Key financial health indicators include liquidity ratios such as the current ratio and quick ratio, which measure a company's ability to meet short-term obligations and provide insights into financial stability and operational efficacy [2]. Profitability ratios like net profit margin, return on assets (ROA), and return on equity (ROE) assess the company's ability to generate profits relative to revenue, assets, and equity, serving as important indicators of financial performance and efficiency. Financial stability ratios, including the debt-to-equity ratio and interest coverage ratio, indicate a company's capacity to meet long-term obligations, offering insights into financial leverage and long-term sustainability [50]. Monitoring techniques such as cash flow analysis are essential for understanding cash inflows and outflows, ensuring liquidity to meet obligations and prevent crises [39], while variance analysis compares actual financial performance with budgeted or forecasted figures, identifying discrepancies and their causes to implement timely corrective actions [30]. Tools for financial health monitoring include financial dashboards, which provide real-time overviews of key metrics using platforms like QuickBooks, Xero, and NetSuite, and financial planning and analysis (FP&A) software like Adaptive Insights and Anaplan, which offer advanced capabilities for planning, budgeting, forecasting, and reporting to maintain financial discipline and enable data-driven decisions. Successful start-ups leveraging financial analytics include Square, which uses financial forecasting and budgeting to manage rapid growth through sophisticated models and scenario analysis [29], Kickstarter, which optimizes crowdfunding campaigns by analyzing donor behavior and campaign performance to enhance fundraising strategies, and Etsy, which monitors financial health using advanced analytics tools to track liquidity, profitability, and solvency ratios, ensuring stability and supporting strategic growth decisions.

13 Utilizing Operational Analytics to Enhance Efficiency and Scalability.

Operational analytics for efficiency and scalability entails employing data analysis to optimize supply chain, logistics, and workforce management, resulting in significant cost reductions, heightened competitive advantage, and higher customer satisfaction for start-ups. Logistics and supply chain analytics optimize the flow of products, services, and information from suppliers to customers, with key applications including demand forecasting, where regression analysis, machine learning models, and time series analysis predict inventory needs to reduce stockouts [36], inventory optimization, which uses techniques like ABC analysis, economic order quantity (EOQ), and safety stock calculations to balance service levels with carrying costs [116], and transportation and logistics optimization, which employs vehicle routing problem (VRP) solutions, real-time monitoring, and linear programming to optimize routes, reduce costs, and improve delivery times [43]. Workforce analytics enhances human resource management and workforce performance by improving employee productivity, engagement, and retention through key applications such as recruitment and talent management, where predictive modeling and machine learning forecast suitable candidates and identify effective recruitment channels [127], performance management, which uses techniques like 360-degree feedback and performance dashboards to monitor and enhance employee performance; and employee engagement and retention, which employs surveys and social network analysis to measure engagement and predictive analytics to identify attrition factors and develop retention strategies [112].

14 Utilizing Operational Analytics to Enhance Efficiency and Scalability

Operational analytics for efficiency and scalability entails employing data analysis to optimize supply chain, logistics, and workforce management, resulting in significant cost reductions, heightened competitive advantage, and higher customer satisfaction for start-ups [126], resource allocation, where data analytics predicts demand and ensures optimal deployment of resources such as workforce scheduling, inventory management, and capacity planning [108], and scalability metrics, where establishing key metrics like transaction volumes, response times, and throughput rates ensures that start-ups can monitor growth and maintain performance standards, with data analytics tracking these metrics for informed decision-making [102]. Successful start-ups leveraging operational analytics include Amazon, which optimizes its inventory management and distribution network using real-time data and predictive analytics to ensure high customer satisfaction and efficient delivery [25], Google, which enhances talent management and improves employee engagement by analyzing data from performance reviews, employee surveys, and social networks to develop effective strategies for recruitment, retention, and employee development [9], and Uber, which optimizes its ride-hailing service scaling by analyzing trip data, driver performance, and customer feedback to improve operations, service quality, and expand market presence [15].

15 Principles and Management of Ethical Behavior in Strategic Analytics

Ethics and governance in strategic analytics entail giving priority to safeguarding data privacy, ensuring security, and addressing ethical concerns in order to protect business operations, adhere to legislation, and uphold customer trust. Data privacy emphasizes personal data protection through policies and procedures that adhere to regulations like GDPR [128], data anonymization techniques such as masking and pseudonymization to minimize privacy risks [33], and consent management to ensure informed user consent [118]. Ensuring data security includes encryption of data in transit and at rest with algorithms like AES for robust protection [87], rigorous access control measures like multi-factor authentication and role-based access control to limit access to sensitive data [110], and regular security audits and compliance reviews to identify vulnerabilities and maintain standards like ISO/IEC 27001 [56]. Ethical considerations in data usage require responsible and transparent use of data, balancing benefits with ethical implications concerning bias, fairness, and accountability. Addressing algorithmic bias involves evaluating and mitigating biases in data collection, model training, and deployment to ensure fairness [4], implementing fairness metrics like demographic parity and equalized odds to measure and ensure predictive model fairness, and inclusive data practices to ensure diversity and representation in data sets [53]. Transparency and accountability in data practices involve clear communication about data collection, use, and dissemination to foster trust among stakeholders [28], and establishing accountability mechanisms like ethics committees, regular audits, and whistleblower protections to monitor and govern data usage effectively [38].

16 Governance Frameworks for Data Management

Governance frameworks for data management ensure that data is managed responsibly, securely, and ethically, establishing standards for data stewardship, quality control, and regulatory compliance. Effective data governance models include data stewardship, where data custodians manage data assets and maintain quality and integrity through the supervision of data policies, standards, and procedures [119], data quality management, which involves processes for data validation, purification, and enrichment to ensure high data quality standards using methodologies like data profiling and data lineage monitoring [91], and regulatory compliance, requiring adherence to data protection regulations such as GDPR and CCPA, with frameworks establishing guidelines for data management, storage, and sharing to ensure compliance [124]. Implementing governance frameworks involves adopting established frameworks and standards like the Data Management Body of Knowledge (DMBOK) and ISO/IEC 38,505–1 for a structured approach [24], establishing data governance committees to oversee data management practices and address ethical considerations [67], and incorporating mechanisms for continuous improvement through regular reviews, audits, and feedback loops to adapt to evolving needs [72]. Case studies highlight ethical and governance challenges in start-ups, such as Facebook, which faced significant challenges related to data privacy and security, exemplified by the Cambridge Analytica scandal that underscored the need for robust data governance structures and greater transparency [60], Google, which has implemented strong data governance frameworks including data stewardship roles and compliance with GDPR to manage its vast data assets responsibly and enhance transparency and accountability [135], and Uber, which has responded to ethical issues related to data privacy and security breaches by strengthening its data governance practices, implementing stricter access controls, enhancing data security measures, and improving transparency [11].

17 Emerging Patterns in Strategic Analytics for Start-Ups

Emerging technologies are constantly reshaping future trends in strategic analytics for start-ups. It is vital to stay updated on these breakthroughs in order to capitalize on new opportunities and maintain a competitive edge. Significant emerging technologies impacting strategic analytics include artificial intelligence and machine learning, with deep learning facilitating more precise predictions and sophisticated data processing capabilities in tasks such as image identification, natural language processing, and autonomous systems [77], and reinforcement learning training algorithms to make a series of decisions based on rewarding good actions, particularly valuable in dynamic settings like real-time bidding and robotics [115]. Blockchain technology enables secure, transparent, and decentralized data management, improving data integrity and trust, especially in industries requiring high levels of data security and traceability [132], while smart contracts, which execute themselves and eliminate the need for intermediaries, lead to cost reduction and increased efficiency [122]. The Internet of Things (IoT) enhances data collection through real-time data from various sources, providing rich datasets for analytics that lead to informed decision-making and improved operational efficiency [3], and IoT analytics can predict equipment failures and schedule maintenance before issues occur, reducing downtime and maintenance costs in manufacturing and logistics [76]. The future of work, shaped by technological advancements and evolving workplace dynamics, sees analytics playing a critical role in understanding these changes and preparing organizations. In remote work and virtual teams, productivity analytics track and measure productivity in remote environments, helping managers optimize workflows and support employee well-being [73], while advanced analytics in collaboration tools provide insights into team dynamics, communication patterns, and project progress to manage virtual teams effectively [92]. Workforce analytics can identify skill gaps and suggest targeted training programs to ensure personnel have the necessary competencies to adapt to evolving workplace demands [6], and provide insights into career progression patterns, helping employees and managers plan effective career paths aligned with organizational goals and personal aspirations.

17.1 Preparing for the Next Wave of Disruption

In order to be ready for the upcoming wave of disruption, start-ups must maintain flexibility and be well-equipped to adapt to technology and market shifts. Strategic analytics play a crucial role in providing the essential foresight and tools. Anticipating market trends involves utilizing analytics to identify and predict trends, employing techniques such as sentiment analysis and market basket analysis to gain insights into consumer behavior and emerging trends [1], and using scenario planning to create specific and realistic scenarios to guide strategic decision-making and prepare for various potential futures [113]. Innovation and agility are fostered through innovation analytics, which pinpoints areas for innovation by analyzing customer feedback, market gaps, and technological advancements, helping start-ups to innovate continuously and stay relevant [19], and implementing agile methodologies supported by analytics, which facilitate rapid iteration and continuous improvement, enabling agile teams to make informed decisions and adapt swiftly to changes [105].

17.2 Analysis of Successful Start-Ups Prepared for the Future

Case studies of forward-thinking start-ups illustrate how smart utilization of analytics has facilitated substantial expansion and a competitive edge. Tesla utilizes cutting-edge technologies such as artificial intelligence (AI), the Internet of Things (IoT), and blockchain to foster innovation in the automotive sector, employing predictive analytics and real-time data monitoring to enhance vehicle performance and customer experience [51]. Airbnb leverages advanced analytics to understand market trends, optimize pricing, and enhance user experience, using machine learning and big data to stay ahead of market disruptions and maintain a competitive edge [46]. Zoom capitalized on the shift to remote work by using analytics to understand user behavior and improve its platform, continuously innovating and scaling operations with data insights to become a leader in virtual collaboration [88]. Specifically, Airbnb employs machine learning algorithms to analyze user data and offer tailored recommendations, using dynamic pricing strategies to adjust rental prices based on demand fluctuations, thereby optimizing revenues and maintaining competitive pricing for guests [46, 131]. Uber uses real-time analytics to optimize driver-rider pairings by analyzing rider location, driver availability, and traffic conditions, and implements surge pricing during high demand periods to dynamically adjust prices based on supply and demand fluctuations, thus improving operational efficiency and satisfaction [15, 22]. Zoom analyzes user engagement data to understand platform usage, employing metrics like meeting durations and feature utilization to continuously improve the product and user experience, and uses analytics to monitor system performance, ensuring scalability and high service quality [88, 123]. These examples underscore how data-driven strategies are critical in driving growth, improving customer satisfaction, and maintaining a competitive edge in dynamic markets.

18 Insights Gained and Optimal Approaches

Valuable knowledge and effective strategies gained from prosperous start-up companies provide essential guidance for utilizing analytics in business operations, including taking a customer-centric approach by utilizing data analytics to deeply understand clients and customize products and services, which enhances customer satisfaction [46, 88], maintaining agility and responsiveness through real-time analytics to promptly adapt to evolving circumstances and meet customer demands, essential for staying competitive and taking advantage of market opportunities [15], efficiently utilizing analytics to facilitate scalable expansion by streamlining processes and assuring a consistent level of service quality, with resources allocated to develop a strong data infrastructure capable of managing growing data volumes and complexity [123], and incorporating data-driven decision-making into processes to ensure plans are grounded in factual evidence rather than intuition, resulting in more informed and impactful corporate decisions [22]. Best practices include investing in analytics capabilities by hiring skilled data scientists and acquiring advanced analytics tools [25], cultivating a data-driven culture by promoting data-based decision-making and continuous learning across the organization with training on data literacy and encouraging data sharing [74], complying with data privacy legislation and establishing strong security measures like data anonymization and encryption to safeguard customer data and uphold trust [128], employing modern analytics approaches such as machine learning, predictive analytics, and real-time analytics for deeper insights and more precise forecasts to improve strategic decision-making [77], and using continuous monitoring and optimization practices to regularly analyze analytics outcomes and adapt plans accordingly, ensuring swift responses to market changes through A/B tests and model adjustments [68].

19 The Road Ahead for Start-Ups in Strategic Analytics

The future prospects for start-ups in strategic analytics entail embracing technical advancements, cultivating a culture centered around data, guaranteeing ethical and responsible utilization of data, and harnessing analytics to drive innovation and expansion. Start-ups can gain a competitive advantage by utilizing technologies such as AI, machine learning, blockchain, and IoT, while continuous learning and adaptation are essential for fully exploiting these advancements [77]. Building a data-driven culture requires investing in data literacy, encouraging cross-functional collaboration, and promoting the use of analytics in all business operations to ensure insights are effectively translated into action [74]. Ensuring ethical and responsible use of data involves prioritizing ethical considerations and implementing robust governance frameworks that include adhering to data privacy regulations, ensuring transparency in data practices, and addressing biases in algorithms to build trust with stakeholders and ensure long-term sustainability [128]. Leveraging analytics for innovation and growth enables start-ups to identify new opportunities, optimize business processes, and scale operations efficiently, establishing themselves as industry leaders through data-driven strategies [19]. Entrepreneurs are called to invest in analytics capabilities by hiring skilled data scientists, acquiring advanced tools, and fostering partnerships with analytics providers, which will improve decision-making and competitive advantage [25]. Cultivating a mindset that prioritizes data-informed decision-making involves promoting the importance of factual decision-making at every company level and providing training and resources to enhance employees' data literacy [74]. In a rapidly changing business environment, staying agile and adaptable is key, continuous monitoring of market trends, customer behavior, and operational performance through analytics allows for strategic pivots based on data insights to stay ahead of the competition [105]. Prioritizing ethics and governance entails committing to ethical data practices, implementing robust governance frameworks, ensuring compliance with regulations, and addressing ethical considerations such as privacy, transparency, and bias, which are crucial for establishing trust with customers and stakeholders and achieving sustained success [128].

References

1. Aggarwal, C.C., Bhuiyan, M.A., Hasan, M.A.: Frequent pattern mining algorithms: a survey. In: Aggarwal, C.C., Han, J. (eds.) Frequent pattern mining, pp. 19–64. Springer (2012)
2. Altman, E.I.: Financial ratios, discriminant analysis and the prediction of corporate bankruptcy. J. Fin. **23**(4), 589–609 (1968)
3. Atzori, L., Iera, A., Morabito, G.: The Internet of Things: a survey. Comput. Netw.. Netw. **54**(15), 2787–2805 (2010)
4. Barocas, S., Selbst, A.D.: Big data's disparate impact. Calif. Law Rev. **104**, 671–732 (2016)
5. Benninga, S.: Financial Modeling. MIT Press (2014)
6. Bersin, J.: A new paradigm for corporate training: learning in the flow of work (2018, June 5). Retrieved from https://joshbersin.com/2018/06/a-new-paradigm-for-corporate-training-learning-in-the-flow-of-work/
7. Bertsimas, D., Kallus, N.: From predictive to prescriptive analytics. Manage. Sci. **66**(3), 1025–1044 (2020)
8. Blank, S.: Why the lean start-up changes everything. Harv. Bus. Rev.. Bus. Rev. **91**(5), 63–72 (2013)
9. Bock, L.: Work Rules!: Insights from Inside Google That Will Transform How You Live and Lead. John Murray (2015)
10. Butterfield, S.: Slack's approach to product development (2015). Retrieved from https://slack.com/blog/product/slack-approach-to-product-development
11. Calo, R., Rosenblat, A.: The taking economy: Uber, information, and power. Columbia Law Rev. **117**(6), 1623–1690 (2017)
12. Camp, R.C.: Benchmarking: The Search for Industry Best Practices that Lead to Superior Performance. Quality Press (1989)
13. Checkland, P.: Systems Thinking. John Wiley & Sons, Systems Practice (1981)
14. Chen, H., Chiang, R.H., Storey, V.C.: Business intelligence and analytics: From big data to big impact. MIS Q. **36**(4), 1165–1188 (2012)
15. Chen, M., Mao, S., Liu, Y.: Big data: a survey. Mobile Netw. Appl. **19**(2), 171–209 (2017)
16. Chaffey, D., Ellis-Chadwick, F.: Digital Marketing: Strategy, Implementation and Practice. Pearson UK (2019)
17. Chambers, J.C., Mullick, S.K., Smith, D.D.: How to choose the right forecasting technique. Harv. Bus. Rev.. Bus. Rev. **49**(4), 45–74 (1971)
18. Chermack, T.J.: Scenario Planning in Organizations: How to Create, Use, and Assess Scenarios. Berrett-Koehler Publishers (2011)
19. Christensen, C.M., Raynor, M.E., McDonald, R.: What is disruptive innovation? Harv. Bus. Rev.. Bus. Rev. **93**(12), 44–53 (2015)
20. Clifton, B.: Advanced Web Metrics with Google Analytics. John Wiley & Sons (2012)

21. Court, D., Elzinga, D., Mulder, S., Vetvik, O.J.: The consumer decision journey. McKinsey Q. **3**(3), 96–107 (2009)
22. Cramer, J., Krueger, A.B.: Disruptive change in the taxi business: the case of Uber. Am. Econ. Rev. **106**(5), 177–182 (2016)
23. Croll, A., Yoskovitz, B.: Lean Analytics: Use Data to Build a Better Startup Faster. O'Reilly Media (2013)
24. DAMA.: The DAMA Dictionary of Data Management. Technics Publications (2017)
25. Davenport, T.H.: Big Data at Work: Dispelling the Myths, Uncovering the Opportunities. Harvard Business Review Press (2014)
26. Davenport, T.H., Bean, R.: Big data and AI executive survey 2018. New Vantage Partners, pp. 1–29 (2018)
28. Diakopoulos, N.: Accountability in algorithmic decision making. Commun. ACM. ACM **59**(2), 56–62 (2016)
29. Dorsey, J.: Square's approach to financial planning (2016). Retrieved from https://squareup.com/about-us/our-company
30. Drury, C.: Management and Cost Accounting. Cengage Learning (2013)
31. Edelman, D.C., Singer, M.: Competing on customer journeys. Harv. Bus. Rev.. Bus. Rev. **93**(11), 88–100 (2015)
32. Eccles, R.G., Herz, R.H., Keegan, E.M., Phillips, D.M.H.: The Value Reporting Revolution: Moving Beyond the Earnings Game. John Wiley & Sons (2001)
33. El Emam, K., Dankar, F.K.: Protecting privacy using k-anonymity. J. Am. Med. Inform. Assoc. **15**(5), 627–637 (2008)
34. Fader, P.S., Hardie, B.G.: Probability models for customer-base analysis. J. Interact. Mark. **23**(1), 61–69 (2009)
35. Few, S.: Information Dashboard Design: Displaying Data for At-a-Glance Monitoring. Analytics Press (2013)
36. Fildes, R., Goodwin, P., Lawrence, M., Nikolopoulos, K.: Effective forecasting and judgmental adjustments: an empirical evaluation and strategies for improvement in supply-chain planning. Int. J. Forecast. **35**(1), 274–287 (2019)
37. Fisher, R.A.: Statistical Methods for Research Workers. Oliver and Boyd (1925)
38. Floridi, L., Cowls, J., Beltrametti, M., Chatila, R., Chazerand, P., Dignum, V., Schafer, B.: AI4People—an ethical framework for a good AI society: opportunities, risks, principles, and recommendations. Mind. Mach. **28**(4), 689–707 (2018)
39. Fraser, L.M., Ormiston, A.: Understanding Financial Statements. Pearson (2016)
40. Friedman, J., Hastie, T., Tibshirani, R.: The Elements of Statistical Learning. Springer Series in Statistics (2001)
41. Gomez-Uribe, C.A., Hunt, N.: The Netflix recommender system: algorithms, business value, and innovation. ACM Trans. Manage. Inf. Syst. (TMIS) **6**(4), 13 (2016)
42. Goodfellow, I., Bengio, Y., Courville, A.: Deep Learning. MIT Press (2016)
43. Govindan, K., Soleimani, H., Kannan, D.: Reverse logistics and closed-loop supply chain: a comprehensive review to explore the future. Eur. J. Oper. Res.Oper. Res. **240**(3), 603–626 (2015)
44. Green, P.E., Krieger, A.M.: Segmenting markets with conjoint analysis. J. Mark. **55**(4), 20–31 (1991)
45. Grolinger, K., Hayes, M., Higashino, W.A., L'Heureux, A., Capretz, M.A.M.: Challenges for mapreduce in big data. In IEEE World Congress on Services, pp. 182–189 (2014)
46. Guttentag, D.: Airbnb: disruptive innovation and the rise of an informal tourism accommodation sector. Curr. Issue Tour.. Issue Tour. **18**(12), 1192–1217 (2015)
47. Gürel, E., Tat, M.: SWOT analysis: a theoretical review. J. Int. Soc. Res. **10**(51), 994–1006 (2017)
48. Hastie, T., Tibshirani, R., Friedman, J.: The Elements of Statistical Learning: Data Mining, Inference, and Prediction. Springer Science & Business Media (2009)
49. Hays, C.L.: Wal-Mart: The Face of Twenty-first-century Capitalism. New Press (2004)
50. Higgins, R.C.: Analysis for Financial Management. McGraw-Hill/Irwin (2012)

51. Higgins, T.: How Tesla is driving itself into the future. The Wall Street Journal (2017). Retrieved from https://www.wsj.com/articles/how-tesla-is-driving-itself-into-the-future-1509549647
52. Hillier, F.S., Lieberman, G.J.: Introduction to Operations Research. McGraw-Hill (2001)
53. Holstein, K., Wortman Vaughan, J., Daumé III, H., Dudik, M., Wallach, H.: Improving fairness in machine learning systems: what do industry practitioners need? In: Proceedings of the 2019 CHI Conference on Human Factors in Computing Systems (pp. 1–16) (2019)
54. Hopkin, P.: Fundamentals of Risk Management: Understanding, Evaluating, and Implementing Effective Risk Management. Kogan Page Publishers (2018)
55. Houston, D.: Dropbox: startup lessons learned (2014). Retrieved from https://www.dropbox.com/startup-lessons-learned
56. Humphreys, E.: Information security management standards: compliance, governance, and risk management. Inf. Secur. Tech. Rep.Secur. Tech. Rep. **13**(4), 247–255 (2008)
57. Hyndman, R.J., Athanasopoulos, G.: Forecasting: Principles and Practice. OTexts (2018)
58. Iansiti, M., Lakhani, K.R.: Managing our hub economy. Harv. Bus. Rev.. Bus. Rev. **95**(5), 84–92 (2017)
59. Inmon, W.H.: Building the Data Warehouse. John Wiley & Sons (2005)
60. Isaak, J., Hanna, M.J.: User data privacy: Facebook, Cambridge Analytica, and privacy protection. Computer **51**(8), 56–59 (2018)
61. James, G., Witten, D., Hastie, T., Tibshirani, R.: An Introduction to Statistical Learning: With Applications in R. Springer (2013)
62. Johnston, M.W., Marshall, G.W.: Sales Force Management: Leadership, Innovation, Technology. Routledge (2016)
63. Jordan, M.I., Mitchell, T.M.: Machine learning: trends, perspectives, and prospects. Science **349**(6245), 255–260 (2015)
64. Kaplan, R.S., Cooper, R.: Cost & Effect: Using Integrated Cost Systems to Drive Profitability and Performance. Harvard Business School Press (1998)
65. Kaplan, R.S., Norton, D.P.: The balanced scorecard: measures that drive performance. Harv. Bus. Rev.. Bus. Rev. **70**(1), 71–79 (1992)
66. Kaufman, L., & Rousseeuw, P.J.: Finding Groups in Data: An Introduction to Cluster Analysis. John Wiley & Sons (2009)
67. Khatri, V., Brown, C.V.: Designing data governance. Commun. ACM. ACM **53**(1), 148–152 (2010)
68. Kohavi, R., Longbotham, R.: Online controlled experiments and A/B testing. In: Sammut, C., Webb, G.I. (eds.) Encyclopedia of Machine Learning and Data Mining, pp. 922–929. Springer (2017)
69. Kotler, P., Keller, K.L.: Marketing Management. Pearson (2016)
70. Kreps, J.: Kafka: A distributed messaging system for log processing. Proc. NetDB **11**, 1–7 (2011)
71. Kujala, S.: User involvement: a review of the benefits and challenges. Behav. Inform. Technol. **22**(1), 1–16 (2003)
72. Ladley, J.: Data Governance: How to Design, Deploy, and Sustain an Effective Data Governance Program. Elsevier (2012)
73. Larson, B.Z., Vroman, S.R., Makarius, E.E.: A guide to managing your (newly) remote workers. Harvard Business Review (2020, March 18). Retrieved from https://hbr.org/2020/03/a-guide-to-managing-your-newly-remote-workers
74. LaValle, S., Lesser, E., Shockley, R., Hopkins, M.S., Kruschwitz, N.: Big data, analytics and the path from insights to value. MIT Sloan Manag. Rev.Manag. Rev. **52**(2), 21–32 (2011)
75. Lazear, E.P.: Balanced skills and entrepreneurship. Am. Econ. Rev. **94**(2), 208–211 (2004)
76. Lee, J., Bagheri, B., Kao, H.A.: A cyber-physical systems architecture for industry 4.0-based manufacturing systems. Manuf. Lett. **3**, 18–23 (2014)
77. LeCun, Y., Bengio, Y., Hinton, G.: Deep learning. Nature **521**(7553), 436–444 (2015)
78. Lemon, K.N., Verhoef, P.C.: Understanding customer experience throughout the customer journey. J. Mark. **80**(6), 69–96 (2016)

79. Lemmens, A., Croux, C.: Bagging and boosting classification trees to predict churn. J. Mark. Res. **43**(2), 276–286 (2006)
80. Li, H., Kannan, P.K.: Attributing conversions in a multichannel online marketing environment: an empirical model and a field experiment. J. Mark. Res. **51**(1), 40–56 (2014)
81. Liu, B.: Sentiment analysis and opinion mining. Synthesis Lect. Human Lang. Technol. **5**(1), 1–167 (2012)
82. Loudon, K.C., Loudon, J.P.: Management Information Systems: Organization and Technology. Prentice Hall (2000)
83. Macal, C.M., North, M.J.: Tutorial on agent-based modelling and simulation. J. Simul. **4**(3), 151–162 (2010)
84. Manning, C.D., Raghavan, P., Schütze, H.: Introduction to Information Retrieval. Cambridge University Press (2008)
85. Manyika, J., Chui, M., Brown, B., Bughin, J., Dobbs, R., Roxburgh, C., Byers, A.H.: Big Data: The Next Frontier for Innovation, Competition, and Productivity. McKinsey Global Institute (2011)
86. McAfee, A., Brynjolfsson, E.: Big data: the management revolution. Harv. Bus. Rev.. Bus. Rev. **90**(10), 60–68 (2012)
87. Menezes, A.J., Van Oorschot, P.C., Vanstone, S.A.: Handbook of Applied Cryptography. CRC Press (2018)
88. Molla, R.: Zoom's explosive growth is slowing down. Here's what comes next. Vox (2020, August 31). Retrieved from https://www.vox.com/recode/2020/8/31/21408457/zoom-explosive-growth-slowing-down-future-video-calls-pandemic
89. Murphy, K.P.: Machine Learning: A Probabilistic Perspective. MIT Press (2012)
90. Nassirtoussi, A.K., Aghabozorgi, S., Wah, T.Y., Ngo, D.C.L.: Text mining for market prediction: a systematic review. Expert Syst. Appl. **41**(16), 7653–7670 (2014)
91. Olson, J.E.: Data Quality: The Accuracy Dimension. Morgan Kaufmann (2003)
92. Olson, G.M., Olson, J.S.: Working together apart: collaboration over the internet. In: Jacko, J.A. (ed.) The Human-Computer Interaction Handbook, pp. 549–564. CRC Press (2012)
93. Pang, B., Lee, L.: Opinion mining and sentiment analysis. Found. Trends® Inform. Retrieval **2**(1–2), 1–135 (2008)
94. Peters, K., Chen, Y., Kaplan, A.M., Ognibeni, B., Pauwels, K.: Social media metrics—a framework and guidelines for managing social media. J. Interact. Mark. **27**(4), 281–298 (2013)
95. Petrillo, F., Penteado, R., Pimenta, M., Dietrich, C.: What went wrong? A survey of problems in agile software development projects. J. Syst. Softw.Softw. **82**(12), 311–324 (2009)
96. Porter, M.E.: Competitive Strategy: Techniques for Analyzing Industries and Competitors. Free Press (1980)
97. Porter, M.E.: Competitive Advantage: Creating and Sustaining Superior Performance. Free Press (1985)
99. Provost, F., Fawcett, T.: Data science for business: what you need to know about data mining and data-analytic thinking. O'Reilly Media, Inc. (2013)
100. Pyhrr, P.A.: Zero-base budgeting. Harv. Bus. Rev.. Bus. Rev. **51**(6), 111–121 (1973)
101. Raghupathi, W., Raghupathi, V.: Big data analytics in healthcare: promise and potential. Health Inform. Sci. Syst. **2**(1), 3 (2014)
102. Rehman, M.H., Ahmed, E., Yaqoob, I., Jayaraman, P.P.: Big data analytics in industrial IoT using a concentric computing model. IEEE Commun. Mag.Commun. Mag. **56**(2), 37–43 (2018)
103. Reichheld, F.F.: The one number you need to grow. Harv. Bus. Rev.. Bus. Rev. **81**(12), 46–55 (2003)
104. Richardson, A.: Using customer journey maps to improve customer experience. Harvard Business Review (2010, November). Retrieved from https://hbr.org/2010/11/using-customer-journey-maps-to
105. Rigby, D.K., Sutherland, J., Takeuchi, H.: Embracing agile. Harv. Bus. Rev.. Bus. Rev. **94**(5), 40–50 (2016)

106. Ries, E.: The lean startup: How today's entrepreneurs use continuous innovation to create radically successful businesses. Crown Business (2011)
107. Rubinstein, R.Y., Kroese, D.P.: Simulation and the Monte Carlo Method. John Wiley & Sons (2016)
108. Sahin, F., Robinson, E.P.: Flow coordination and information sharing in supply chains: review, implications, and directions for future research. Decis. Sci.. Sci. **33**(4), 505–536 (2002)
109. Sallam, R.L., Richardson, J., Hagerty, J., Hostmann, B.: Magic quadrant for business intelligence and analytics platforms. Gartner RAS Core Research Note G, 282092 (2014)
110. Sandhu, R.S., Coyne, E.J., Feinstein, H.L., Youman, C.E.: Role-based access control models. IEEE Comp. **29**(2), 38–47 (1996)
111. Sargeant, A., Jay, E.: Fundraising Management: Analysis, Planning and Practice. Routledge (2004)
112. Schiemann, W.A.: Beyond engagement: how to make your business an idea factory. Society for Human Resource Management (2014)
113. Schoemaker, P.J.: Scenario planning: a tool for strategic thinking. Sloan Manage. Rev. **36**(2), 25–40 (1995)
114. Siegel, E.: Predictive analytics: the power to predict who will click, buy, lie, or die. Wiley (2016)
115. Silver, D., Huang, A., Maddison, C.J., Guez, A., Sifre, L., Van Den Driessche, G., Hassabis, D.: Mastering the game of Go with deep neural networks and tree search. Nature **529**(7587), 484–489 (2016)
116. Silver, E.A., Pyke, D.F., Thomas, D.J.: Inventory management and production planning and scheduling. John Wiley & Sons (1998)
117. Singh, V., Venugopal, P., Manjaly, J.A.: Sales and distribution management: a practice-based approach. SAGE Publications India (2015)
118. Solove, D.J.: Privacy self-management and the consent dilemma. Harv. Law Rev.. Law Rev. **126**, 1880–1903 (2013)
119. Soares, S.: The chief data officer handbook for data governance. MC Press (2015)
120. Stonebraker, M.: SQL databases v. NoSQL databases. Commun. ACM. ACM **53**(4), 10–11 (2018)
121. Sutton, R.S., Barto, A.G.: Reinforcement learning: an introduction. MIT Press (2018)
122. Szabo, N.: The idea of smart contracts. Nick Szabo's Papers and Concise Tutorials (1997)
123. Tang, H., Abdrashitov, M., Raj, V.: Meeting the challenges of operating Zoom at scale. Zoom Blog (2020). Retrieved from https://blog.zoom.us/meeting-the-challenges-of-operating-zoom-at-scale/
124. Tene, O., Polonetsky, J.: Big data for all: privacy and user control in the age of analytics. Northwest. J. Technol. Intellect. Prop. **11**(5), 239–273 (2013)
125. Turban, E., Sharda, R., Delen, D., King, D.: Business intelligence: a managerial approach. Prentice Hall (2011)
126. Van der Aalst, W.: Process mining: data science in action. Springer (2016)
127. Van Iddekinge, C.H., Eidson, C.E., Roth, P.L.: A meta-analysis of selection interview reliability and validity: a follow-up. J. Appl. Psychol. **101**(3), 599–619 (2016)
128. Voigt, P., Von dem Bussche, A.: The EU general data protection regulation (GDPR): a practical guide. Springer (2017)
129. Vonderau, P.: The Spotify effect: digital distribution and financial growth. Telev. New Media. New Media **20**(1), 3–19 (2019)
130. Wamba, S.F., Akter, S., Edwards, A., Chopin, G., Gnanzou, D.: How 'big data' can make big impact: findings from a systematic review and a longitudinal case study. Int. J. Prod. Econ. **165**, 234–246 (2015)
131. Ye, S., Qiu, Y., Huang, T., Li, Y.: Dynamic pricing in the sharing economy: evidence from Airbnb. Manage. Sci. **64**(12), 5884–5898 (2018)
132. Yli-Huumo, J., Ko, D., Choi, S., Park, S., Smolander, K.: Where is current research on blockchain technology? A systematic review. PLoS ONE **11**(10), e0163477 (2016)

134. Zikopoulos, P.C., Eaton, C., DeRoos, D., Deutsch, T., Lapis, G.: Understanding big data: analytics for enterprise class Hadoop and streaming data. McGraw-Hill Osborne Media (2011)
135. Zuboff, S.: The age of surveillance capitalism: the fight for a human future at the new frontier of power. PublicAffairs (2019)

Optimization Techniques for Strategic Analytics

Gurwinder Singh and Amarinder Singh

Abstract This chapter discusses on various optimization approaches that can be used strategically in decision making areas. Therefore it underlines the importance of these techniques in the improvement of the supply chain's performance and sustainability. When it comes to the modern realities of doing business, strategic analytics are one of the key tools for decision-making that increases companies' adaptation to their environments. The chapter helps to understand the state-of-practice approaches and identifies the potential of using data analytical methods and presents the performances of multi-objective optimisation methods including evolutionary algorithms, swarm intelligence, hybrid metaheuristic and hyper heuristic methods. The chapter begins with definitions of descriptive, predictive, and prescriptive analytics with fundamental concepts. The latter offers an analysis of how the Particle Swarm Optimization (PSO) mathematical algorithm can be applied to improve a specific logistics schedule. It mainly contains problem definition, data collection, determination of the fitness function and PSO algorithm to determine the best routes. Different configuration of PSO is shown in the context of the case study to understand how it influences optimization results such as cost, transit time and computation time. Thus, this paper aims to focus on stressing tangible benefits, which strategic analytics brings, including increased productivity, competitive advantage, optimization of supply networks and chain and enhancements in financial outcomes. It is always wise to learn from experience and from others and this chapter will be of immense value to any business or student and the guidelines will be very helpful in the implementation of optimization techniques in improving business resilience.

Keywords Optimization techniques · Strategic analytics · Swarm intelligence · Metaheuristics · Logistics optimization · Multi-objective optimization · Data-driven decision-making · Computational efficiency · Predictive analytics · Descriptive analytics · Sustainability in supply chain

G. Singh (✉)
Department of AIT-CSE, Chandigarh University, Mohali, Punjab, India
e-mail: singh1001maths@gmail.com

A. Singh
Chitkara University IET, Chitkara University, Rajpura, Punjab, India

S. Kautish et al. (eds.), *Strategy Analytics for Business Resilience Theories and Practices*, Information Systems Engineering and Management 33,
https://doi.org/10.1007/978-3-031-82369-5_9

1 Introduction

Strategic analytics can be defined as the application of specific, complex analytical tools, that aid and underpin strategic decision making processes based in organizations. This approach involves working with data, derivative methods like statistics and machine learning, and mathematical "finding the best solutions" such as optimisation algorithms in order to know more, predict new tendencies and make sensible decisions in accordance to the strategic plan of the company.

Strategic analytics can be described as a group of techniques that can be used to understand complicated information. These methodologies include descriptive analysis for explorative analysis of past activities, predictive analysis for the analysis of future activities based on statistical models and machine learning and prescriptive analysis for making recommendations based on the outcomes of the predictions made. Altogether, these techniques allow organizations to view the tendencies in the past, predict the tendencies in the future, and make sound decisions. The principal purpose of carrying out this analysis is to enable the achievement of accurate and reliable strategies as well as making strategic decisions.

This also helps in the performance of the scenario analysis so that the organizations can assess diverse plans and to control the risks involved. An important component of this process involves data aggregation and understanding that entails the grouping of extensive data sets derived from several sources culminating in arriving at predictive patterns that may be valuable. Optimisation also therefore contributes to the profile of strategic analytics because it offers methods for choosing the best solution given certain boundaries for restrains. For instance, they can determine when to transport goods in a supply chain in order to avoid traffic jam or when to mix the recommended portfolio in the financial service sector in order to meet a certain strategic organisational goals and objectives in the shortest possible time.

Several methods for multi-objective optimisation have been proposed and implemented in different industries to troubleshoot logistics schedules, and its possibilities for enhancement of scenarios of the supply chain are quite significant.

Evolutionary techniques are based on the theory of Darwinian evolution and genetics; in fact, they are often used in multi-objective optimization tasks. For example, Ehtesham Rasi and Sohanian have applied the genetic algorithm to a sustainable supply chain to demonstrate that the EAs can be useful in solving multi-objective problems [1]. Karimi-Mamaghan et al. have addressed a combination of a learning based meta-heuristic with GA for planning inspection under uncertainty in agile environments [2]. Fathollahi-Fard et al. applied a novel and efficient meta-model with multiple objectives for sustainable harvesting and have developed an efficient multi-objective meta-heuristic model in the context of agricultural supply chain [3]. Swarm intelligence is based on the principles similar to those that can be observed in complete organization of ants or flocks of birds. For instance, Moncayo-Martínez and Zhang have presented a multi-objective ant colony optimization for supply chain design good for variable logistics contexts [4].

Hybrid meta-heuristics combine multiple optimization techniques to leverage the strengths of each and mitigate their weaknesses. For instance, Talbi has reviewed hybrid meta-heuristics for multi-objective optimization applications [5]. Bezoui et al. have discussed the hybrid meta-heuristics for supply chain optimization [6]. Seydanlou et al. have proposed a multi-criteria optimisation framework for a closed-loop supply chain network in the olive industry [7]. Momenitabar et al. have combined machine learning with meta-heuristic to develop a sustainable bio-ethanol supply chain network [8]. On the other hand, hyper-heuristics are advanced techniques for choosing or creating heuristics to address computational search problem. Mahmud et al. have introduced a self-adaptive hyper-heuristic for multi-objective optimization approach for integrated supply chain scheduling problems, adaptive in dynamic environments [9].

Other meta-heuristic approaches have addressed specific challenges in supply chain optimization. Farmand et al. have successfully resolved a multi-objective supply chain scheduling using two meta-heuristic algorithms (simulated annealing and tabu search), effective in parallel machines environments [10]. Faramarzi-Oghani et al. have reviewed meta-heuristics for sustainable supply chain management [11]. Coelho et al. have reviewed multi-objective evolutionary algorithms and meta-heuristics for feature selection, bridging the gap between supply chain optimization and data science [12]. Amodeo et al. have contrasted several meta-heuristic methods for supply chain inventory management multi-objective simulation-based optimisation [13]. These studies integrate sustainability and uncertainty into multi-objective optimization techniques for specialized supply chains. Canales-Bustos et al. have developed a de-carbonized mining supply chain [14]. Elfarouk et al. have optimized stochastic, sustainable closed-loop supply chains [15]. Gholian-Jouybari et al. have explored it for sustainable agri-food supply chains [16]. Momenitabar et al. have combined machine learning and meta-heuristics for a bioethanol supply chain network [8]. Ala et al. have introduced a fuzzy multi-objective model for healthcare supply chains [17].

These studies demonstrate the adaptability and efficacy of multi-objective optimisation strategies in improving supply chain efficiency and sustainability across industries. By using modern analytical methodologies, these approaches solve difficult challenges and contribute to strategic decision-making in logistics and beyond.

2 Optimization Techniques

Optimisation techniques are still problem-solving approaches that involve identification of the most appropriate solution for a problem under given constraints The undercover optimization's techniques however are mathematical approaches for arriving at the optimum or the best solution of an issue under given limitations. All these techniques are important in strategic analytics because it is enable organisations to allocate resources, reduce cost and make intelligent decisions Table 1. In the fol-

Table 1 Evolution of optimization techniques

Decade	Optimization technique	Description	Examples
1950s–1960s	Developing stochastic optimization algorithms	Initial development of algorithms that use randomness to find solutions	Random search, Monte Carlo methods
1960s–1970s	Evolutionary algorithms	Algorithms inspired by biological evolution processes like selection, crossover, and mutation	Genetic algorithms, evolution strategies
1980s	Genetic algorithms	Algorithms based on the mechanics of natural genetics and selection	Standard genetic algorithm (SGA), micro-GA
1990s	Swarm intelligence	Algorithms inspired by the collective behavior of decentralized, self-organized systems	Particle swarm optimization (PSO), ant colony optimization (ACO)
1990s–2000s	Local search-based meta-heuristics	Techniques that explore the solution space by iteratively moving to neighbouring solutions	Tabu search, simulated annealing
2000s–present	Hybrid and multi-objective meta-heuristics	Combination of different optimization techniques to solve complex problems and optimize multiple objectives simultaneously	NSGA-II (non-dominated sorting genetic algorithm II), MOEA/D (multi-objective evolutionary algorithm based on decomposition)

lowing subtopics, an analysis of the different significant optimisation methods such as PSO, GA, SA, DE and others like Basin-hopping and Local Search Methods is carried out.

Particle Swarm Optimization (PSO)

Particle Swarm Optimization (PSO) belongs to a class of computational methods that follow the concept of population searching, the inspiration being the social behaviours of flocking birds or schooling fish. Every potential solution is referred as a 'particle,' which traverses through the solution space with the help of current optimum particles. The algorithm modifies the position of the particle based on its own experience and on the experience of other particles of the same swarm.

Algorithm 1: Particle Swarm Optimization (PSO)

Input: Initialize particles randomly

while *stopping criterion not met* **do**
  for *each particle* **do**
    Evaluate the fitness of each particle;
    if *current position is better than personal best* **then**
      Update the personal best position;
    end
    if *current position is better than global best* **then**
      Update the global best position;
    end
  end
  for *each particle* **do**
    Update the velocity and position based on the personal best and global best positions;

$$v_i(t+1) = wv_i(t) + c_1r_1(p_i - x_i(t)) + c_2r_2(g - x_i(t))$$

$$x_i(t+1) = x_i(t) + v_i(t+1)$$

  end
end

where $v_i(t)$ represents the velocity of particle i at time t, $x_i(t)$ the position of particle. Then the p_i depicts its personal best position and g the global best position. Also the inertia-weight w and c_1, c_2 being the acceleration coefficients play an important role for the working of PSO.

Applications in strategic analytics include optimizing routes and schedules for logistics and transportation in supply chain optimization, efficiently allocating resources in project management, and balancing risk and return in financial portfolios through portfolio optimization. In addition, there are various other optimization methods that can be applied in strategic analytics and these include; Genetic Algorithm (GA), Simulated Annealing (SA), and Differential Evolution (DE) while others are Basin-hopping and Local Search. GA is used on a population of solutions with selection, crossover, and mutation to seek better solutions; applied to scheduling, marketing, and network construction. SA mimics the annealing process in metallurgy, probabilistically accepting poor solutions to escape from local optima, and is used in facility location and resource scheduling. DE improves candidate solutions by combining differences between randomly selected solutions, applicable in process optimization and energy management. Basin-hopping and Local Search methods, focusing on escaping local minima and iteratively improving solutions, are also highlighted for their roles in global optimization problems.

3 Case Study: Optimizing Logistics Schedule

This case study aims to optimize the logistics schedule using Particle Swarm Optimization (PSO) based on route availability from Monday to Sunday. The process involved loading the logistics dataset, normalizing costs, handling missing values, and creating a binary availability matrix. The fitness function evaluated route schedules by considering total costs and penalties for unavailable routes. PSO iteratively improved route selection to minimize costs while ensuring availability, resulting in an optimized schedule highlighting cost-effective routes.

3.1 *Problem Formulation*

The Logistics dataset holds data about the existing numbers of routes for transportation. Every route comes with a cost and schedules depend on the day of the week. The dataset includes the following columns: Route ID, Fixed TCO, Port/Airport /Rail Dealing Cost, Fuel Cost, Documentation Requirement Cost, Transit Time (in Hrs.), and Seven days availability (Boolean variables: available on Monday, Tuesday, Wednesday, Thursday, Friday, Saturday or Sunday).

The data preparation steps involve:

1. Loading the dataset from the provided Excel file.
2. Extracting relevant data, such as cost, transit time, and availability columns for each route.
3. Normalizing costs, if necessary, to ensure they are on a comparable scale.
4. Handling any missing values in the dataset.
5. Creating a binary matrix indicating the availability of each route on different days of the week.

3.2 *Fitness Function*

The fitness function evaluates the quality of a given route schedule based on the total cost and availability constraints. It is defined as follows:

$$\text{Fitness} = \text{Total Cost} + \text{Penalty}$$

Total Cost is the sum of all costs associated with the selected routes, calculated as:

$$\text{Total Cost} = \sum_{i=1}^{n} (\text{Fixed Freight Cost}_i + \text{Handling Cost}_i + \text{Fuel Cost}_i + \text{Documentation Cost}_i) \times x_i$$

where x_i is a binary decision variable indicating whether route i is selected (1) or not (0).

Penalty is an additional cost added if the selected routes do not cover all days of the week, calculated as:

$$\text{Penalty} = \sum_{d=\text{Monday}}^{\text{Sunday}} \left(1000 \times \left(1 - \max_i(\text{Availability}_{i,d} \times x_i)\right)\right)$$

This penalty ensures that a large cost is added if any day of the week is not covered by the selected routes. The value 1000 is an arbitrary large number chosen to heavily penalize incomplete schedules.

3.3 Parameters Setting of PSO Algorithm

Parameter Initialization: Initially, after calling the function, the parameters for the swarm size, maximum iteration, inertia weight, cognitive factor, and social factor must be defined. The position and velocity of particles can be initialised randomly within the feasible solution space.

PSO Execution: Begin by evaluating the fitness of each particle using the defined fitness function. Update the personal best and global best positions based on fitness evaluations. Adjust the velocity and position of each particle according to the PSO update equations. Continue this process until the stopping criterion, such as maximum iterations or convergence, is met.

3.4 Implementation and Monitoring

Result Analysis: Analyze the optimized logistics schedule obtained from the PSO algorithm. Compare the optimized schedule with the initial schedule in terms of total costs and route availability. Identify patterns and insights from the optimized route selection to understand cost-effective routes and days of operation.

Validation: Validate the optimized schedule against real-world constraints and operational feasibility. Conduct sensitivity analysis to assess the robustness of the optimized schedule under different cost scenarios and route availabilities.

4 Results and Discussion

Table 2 presents an optimized logistics schedule based on route availability from Monday to Sunday. The table includes various routes, detailing their source and destination, container size, carrier, travel mode, fixed freight cost, handling costs, fuel costs, documentation costs, handling times, and transit times. Each route is evaluated for its availability throughout the week. For instance, routes between Shanghai Railway Station and Wuxi Railway Station via rail have a transit time of 36 h and are available every day. On the other hand, multiple routes involving truck travel, such as those between Singapore Port and Singapore Warehouse, exhibit shorter transit times of around 2–3 h and are also available daily. The detailed costs and transit times

Table 2 Optimized logistics schedule: route details

Route number	Source	Destination	Container size	Carrier	Travel mode
18	Shanghai Railway Station	Wuxi Railway Station	34	Railway authority	Rail
19	Wuxi Railway Station	Shanghai Railway Station	34	Railway authority	Rail
23	Singapore Port	Singapore Warehouse	34	KTM	Truck
24	Singapore Warehouse	Singapore Airport	34	KTM	Truck
26	Singapore Warehouse	Singapore Railway Station	34	KTM	Truck
29	Malaysia Port	Malaysia Warehouse	34	KTM	Truck
31	Malaysia Airport	Malaysia Warehouse	34	KTM	Truck
36	Wuxi Warehouse	Wuxi Airport	34	KTM	Truck
38	Wuxi Warehouse	Wuxi Railway Station	34	KTM	Truck
39	Wuxi Railway Station	Wuxi Warehouse	34	KTM	Truck
41	Shanghai Port	Shanghai Warehouse	34	KTM	Truck
45	Shanghai Railway Station	Shanghai Warehouse	34	KTM	Truck
48	Shanghai Warehouse	Wuxi Warehouse	34	KTM	Truck

Table 3 Optimized logistics schedule: costs

Route number	Fixed freight cost	Port/airport/rail handling cost	Bunker/fuel cost	Documentation cost	Handling time (h)	Extra time (h)
18	80	100	100	50	0	0
19	80	100	100	50	0	0
23	100	0	0	0	0	0
24	100	0	0	0	0	0
26	100	0	0	0	0	0
29	80	0	0	0	0	0
31	80	0	0	0	0	0
36	80	0	0	0	0	0
38	80	0	0	0	0	0
39	80	0	0	0	0	0
41	100	0	0	0	0	0
45	100	0	0	0	0	0
48	125	0	0	0	0	0

provided for each route help in selecting the optimal routes to minimize overall costs while ensuring daily availability.

Table 3 provides detailed information on the costs associated with each route. It includes fixed freight costs, handling costs at ports, airports, or rail stations, bunker or fuel costs, documentation costs, handling times, and any extra times. For example, routes 18 and 19, which are rail routes between Shanghai Railway Station and Wuxi Railway Station, have a fixed freight cost of 80, handling cost of 100, bunker/fuel cost of 100, and documentation cost of 50. These routes do not incur any extra time.

Table 4 details the availability of each route throughout the week. The transit times are provided alongside the days of the week (Monday to Sunday) when the routes are operational. For instance, routes 18 and 19 have a transit time of 36 h and are available every day of the week. In contrast, routes like 23, 24, and 26, which involve truck travel between various Singapore locations, have much shorter transit times of 2 h and are also available daily.

The chart above illustrates the availability of various routes throughout the week. Each horizontal line represents a different route, labeled by its route number on the vertical axis. The days of the week are indicated on the horizontal axis, from Monday to Sunday. The presence of a colored segment on a particular day indicates that the corresponding route is available on that day.

Figure 1 points out the availability of all the logistics routes for the week. Each horizontal line implies one route which is identified on the y-axis with the numerical indicating the route number. The numbering of the days of the week is presented on the abscissa or horizontal axis from Monday to Sunday. The colored segment within the calendar means that the specified particular route is offered on such a day. This chart also gives the management a clear illustration of the different routes that are in operation in a given day and thus helps in planning the long and short-term schedules of the logistics. Routes involving the numbers 30 and 40 are available daily, implying

Table 4 Optimized logistics schedule: availability

Route number	Transit time (h)	Monday	Tuesday	Wednesday	Thursday	Friday	Saturday	Sunday
18	36.0	1	1	1	1	1	1	1
19	36.0	1	1	1	1	1	1	1
23	2.0	1	1	1	1	1	1	1
24	2.0	1	1	1	1	1	1	1
26	2.0	1	1	1	1	1	1	1
29	3.0	1	1	1	1	1	1	1
31	3.0	1	1	1	1	1	1	1
36	3.0	1	1	1	1	1	1	1
38	3.0	1	1	1	1	1	1	1
39	3.0	1	1	1	1	1	1	1
41	3.0	1	1	1	1	1	1	1
45	3.0	1	1	1	1	1	1	1
48	4.0	1	1	1	1	1	1	1

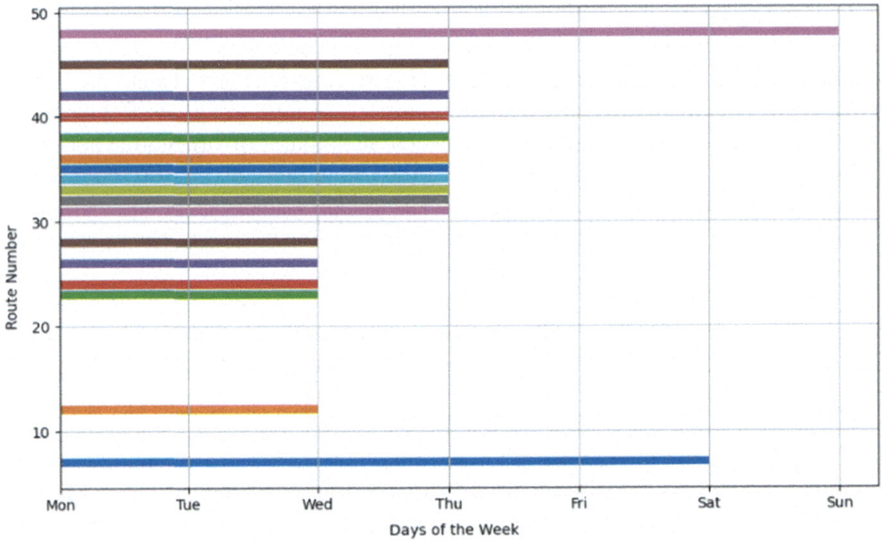

Fig. 1 Route availability throughout the week

high availability and always operating, while routes involving 10 are very scarce and possess low availability.

The four subfigures in Fig. 2 depicts the working of PSO in order to trace the best routes for transportation network. In one of the subfigures in the top left-hand corner, the start state reveals several starting points few as Wuxi Port and Shanghai Port as

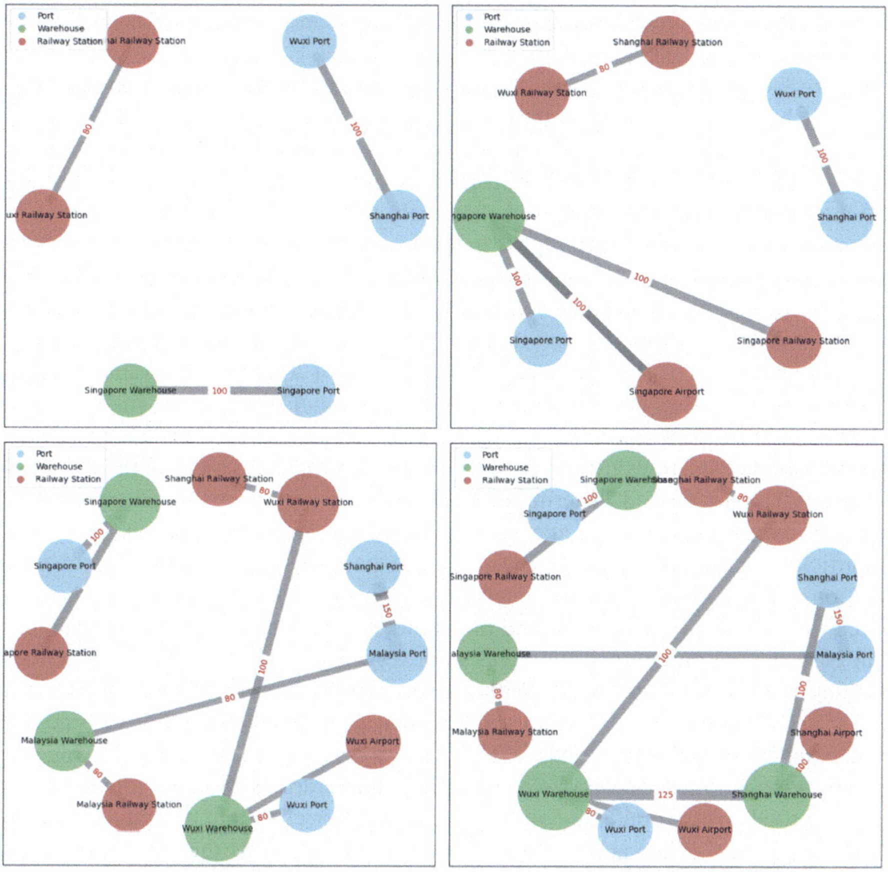

Fig. 2 Progression of the PSO showing the network of selected routes from initial to final state

well as Singapore Warehouse and Singapore Port. As the PSO algorithm continues there is a more advanced stage represented in the top-right subfigure, where more routes are considered and chosen, including the ones like Singapore Warehouse → Singapore Airport, Singapore Warehouse → Singapore Railway Station. Moving forward in the optimization process, there is another big subfigure in the bottom-left corner, which depicts the intermediate form of the network that becomes more extensive with more complex connections. This stage contains such links as between Wuxi Warehouse and Wuxi Railway Station and between Malaysia Warehouse and Malaysia Railway Station. Last of all, the last subfigure in the bottom right is for the PSO optimization of the system. This complete network shows the chosen fundamental pairs by the algorithm that can minimize the costs while maximizing the connection as much as possible. These plans are for example, the route between Singapore Warehouse and Malaysia Port and between Shanghai Railway Station and

Shanghai Port; this shows the details of the extensive and efficient network connection decided by the PSO algorithm.

Figure 3, one can observe how the chosen Particle Swarm Optimization (PSO) algorithm behaves depending on the chosen swarm size and the number of iterations, presenting the dependence of the total cost on the transit time of the schedule of the optimal logistics achieved. From the first plot for the swarm size of 50 and 100 iterations, it has been observed that each case indicates a larger variance of solutions which are very much higher in terms of cost and transit time as compared to other cases of swarm size and number of iterations. The middle plot, with 100 and 200 swarm size, can be observed that the optimization results could be enhanced by enhancing the swarm size and iterations but the costs and transit time of some solutions are still high. The third plot with swarm size of 200 and 400 iterations makes it clear that increasing both swarm size and the number of iterations improves the solution and results in solutions that appear to be more clustered and characterized by lower total costs, transit times that are also lower, meaning more optimal.

Table 5 is summarized in terms of total cost, mean time of transit, number of routes chosen, cost per unit time, time to converge and penalty costs in respect to Particle Swarm Optimization (PSO) with respect to swarm size and number of iterations. The plausible results obtained in this investigation illustrated that the increase in, the swarm size to 50 and the number of iterations to 100 lead to higher total cost and average transit time and a convergence time of 0.32 s and penalties reaching the millions of dollars. With proven conclusions, increasing swarm size to 100 and iterations to 200 improved the total cost and average transit time but deteriorated the penalty costs and reached the convergence time of 1.31 s. The biggest configuration with the swarm size equal to 200 and iterations: 400, has all the lowest values of total cost, average time of transit, and penalty costs, but the longest time to converge—5.46 s. When applied to PSO they help to illustrate the balance between the computational complexity and the performance of the optimisation which in turn assists in identifying the best PSO configuration for the logistics schedules.

Table 6 provides a comparison of optimization results for different swarm sizes and iterations. With a swarm size of 50 and 100 iterations, the total cost is 3595, the average transit time is 14.12 h, and the cost per unit time is 14.98, with a convergence time of 0.33 s and penalty costs of 423.42. For a swarm size of 100 and 200 iterations, the total cost decreases to 1455, with an average transit time of 5.31 h and a cost per unit time of 21.09, converging in 0.78 s but incurring higher penalty costs of 654.84. When the swarm size is increased to 200 with 400 iterations, the total cost is slightly higher at 1525, the average transit time is 5.09 h, and the cost per unit time is 20.61. However, the convergence time significantly increases to 3.53 s, and penalty costs drop to 193.38. This analysis highlights the trade-offs between different optimization configurations in terms of cost, efficiency, and computational time (Fig. 4).

The bar charts illustrate the comparison of total costs and penalty costs across different optimization configurations of swarm size and iterations. The first chart on the left shows that the configuration with a swarm size of 50 and 100 iterations results in the highest total cost of approximately 3595. Larger swarm sizes and more iterations significantly reduce total costs, with configurations of 100, 200 and 200,

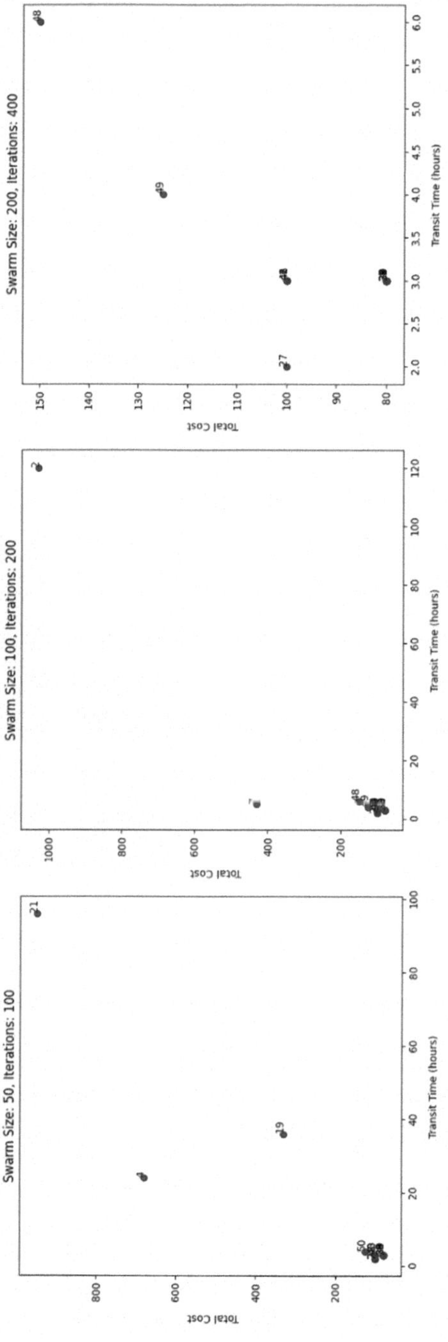

Fig. 3 PSO optimization results showing total cost versus transit time for different swarm sizes and iterations

Table 5 Optimization results for different swarm sizes and iterations

Swarm size	Iterations	Total cost	Avg transit time	Routes	Cost/time	Convergence time (s)	Penalty costs
50	100	4575	11.125	16	25.70	0.32	726.56
100	200	2215	5.12	17	25.46	1.31	839.27
200	400	1230	3.08	13	30.75	5.46	8.90

Table 6 Comparison of swarm sizes and iterations

Total cost	Avg transit time	Routes	Cost	Convergence time (s)	Penalty costs	Swarm size	Max iterations
3595	14.117647	17	14.979167	0.325450	423.420667	50	100
1455	5.307692	13	21.086957	0.779419	654.836594	100	200
1525	5.092308	13	20.608108	3.534611	193.377041	200	400

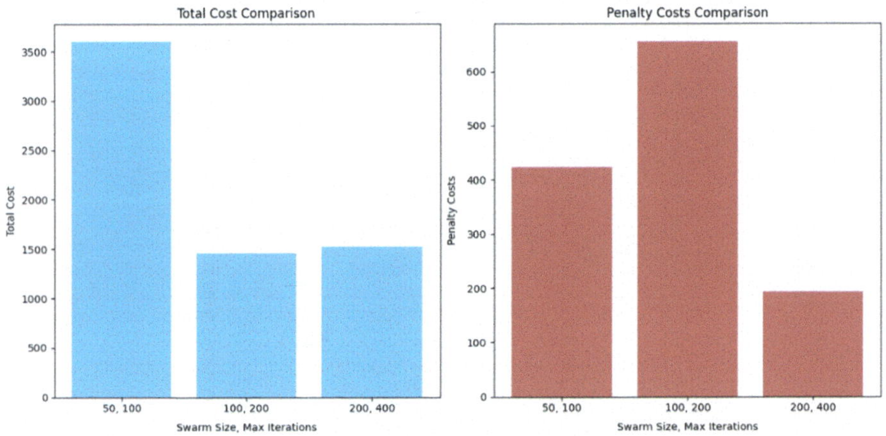

Fig. 4 Comparison of total costs and penalty costs for different swarm sizes and maximum iterations

400 achieving costs of around 1455 and 1525, respectively. However, while the 100, 200 configuration incurs the highest penalty cost of approximately 655, the 200, 400 configuration has the lowest at around 193, demonstrating trade-offs between cost reduction and penalty optimization (Fig. 5).

The plots show that smaller swarm sizes (50 particles) and fewer iterations (100) lead to quick convergence but higher objective function values, indicating less thorough exploration. Increasing the swarm size to 100 particles and iterations to 200 improves results with lower objective function values, while the largest configuration (200 particles, 400 iterations) achieves the best optimization with the lowest objective function value, despite slower initial convergence (Fig. 6).

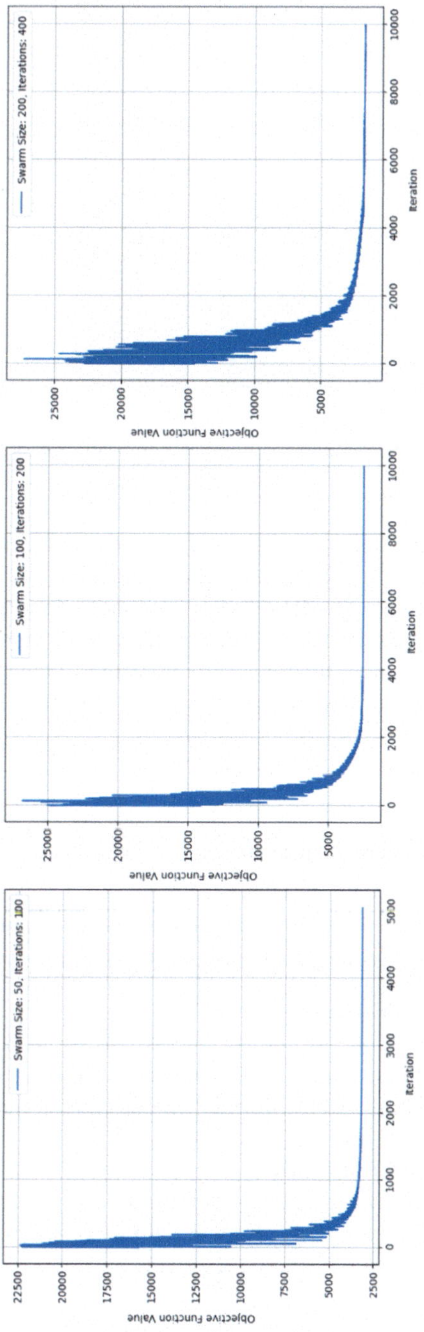

Fig. 5 Objective function value over iterations for different swarm sizes and maximum iterations

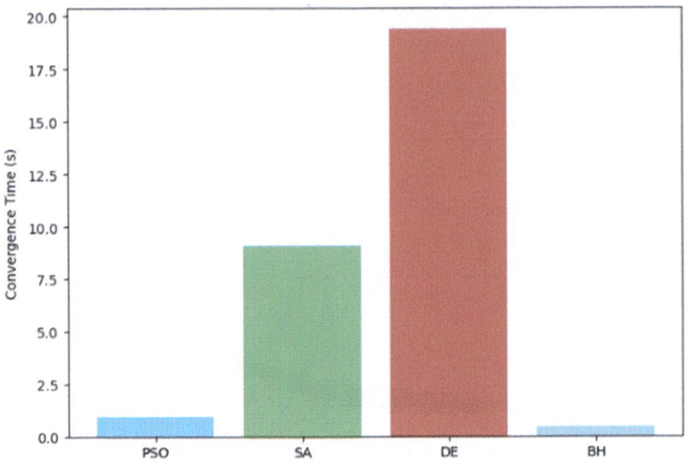

Fig. 6 Comparison of convergence times for different optimization algorithms

The bar chart compares the convergence times of four optimization algorithms: Particle Swarm Optimization (PSO), Simulated Annealing (SA), Differential Evolution (DE), and Basin-hopping (BH). The results show that PSO and BH have the shortest convergence times, indicating their efficiency in quickly finding optimal solutions. Conversely, DE takes the longest to converge, suggesting it requires more computational resources and time. SA's convergence time falls between that of PSO and BH, and DE.

5 Benefits of Strategic Analytics

Strategic analytics provides numerous advantages to organizations, helping them to achieve their goals more effectively and efficiently. Here are some key benefits:

- **Data-Driven Decision-Making**
  Using strategic analytics, organizations can make decisions grounded in solid data rather than intuition or guesswork. Analyzing historical data and identifying trends enables companies to make well-informed choices that are more likely to lead to successful outcomes.
- **Boosted Efficiency and Productivity**
  Strategic analytics optimizes processes and resource allocation, helping organizations operate more efficiently. This results in significant cost savings and increased productivity by using resources more effectively.
- **Competitive Edge**
  Leveraging strategic analytics gives organizations a competitive advantage. By

identifying emerging trends and market opportunities, companies can act quickly to capitalize on these insights and stay ahead of the competition.

- **Optimized Supply Chain and Logistics**
  Strategic analytics enhances supply chain management through lead time reduction and inventory optimisation, and improving delivery performance. This leads to lower costs, increased efficiency, and better customer service.
- **Improved Financial Performance**
  Data-driven decisions help organizations improve their financial performance. Strategic analytics identifies cost-saving opportunities, optimizes pricing strategies, and enhances revenue management.
- **Strategic Alignment and Goal Achievement**
  Strategic analytics ensures that business activities align with the organization's strategic goals. Continuous performance monitoring and data-driven adjustments help organizations to achieve their long-term objectives.

6 Conclusion

This chapter provides a comprehensive overview of various optimization methods and their strategic applications in business analytics, including evolutionary algorithms, swarm intelligence, hybrid meta-heuristics, and hyper-heuristics. The case study on optimizing a logistics schedule using Particle Swarm Optimization (PSO) illustrates practical applications, showcasing detailed steps like problem formulation, data preparation, and fitness function definition to achieve optimal route selection. The results highlight the impact of different PSO configurations on performance metrics such as total cost, transit time, and computational efficiency, demonstrating the strategic analytics' capacity to enhance operational effectiveness, promote well-informed decision-making, and provide businesses a competitive edge. Integrating these optimization techniques into strategic analytics is crucial for modern business resilience, enabling organizations to anticipate trends, mitigate risks, optimize resources, and enhance overall performance.

References

1. Ehtesham Rasi, R., Sohanian, M.: A multi-objective optimization model for sustainable supply chain network with using genetic algorithm. J. Model. Manag. **16**(2), 714–727 (2021)
2. Karimi-Mamaghan, M., Mohammadi, M., Jula, P., Pirayesh, A., Ahmadi, H.: A learning-based metaheuristic for a multi-objective agile inspection planning model under uncertainty. Eur. J. Oper. Res. **285**(2), 513–537 (2020)
3. Fathollahi-Fard, A.M., Tian, G., Ke, H., Fu, Y., Wong, K.Y.: Efficient multi-objective metaheuristic algorithm for sustainable harvest planning problem. Comput. Oper. Res. **158**, 106304 (2023)

4. Moncayo-Martínez, L.A., Zhang, D.Z.: Multi-objective ant colony optimisation: a metaheuristic approach to supply chain design. Int. J. Prod. Econ. **131**(1), 407–420 (2011)
5. Talbi, E.-G.: Hybrid metaheuristics for multi-objective optimization. J. Alg. Comput. Technol. **9**(1), 41–63 (2015)
6. Bezoui, M., Almaktoom, A.T., Bounceur, A., Qaisar, S.M., Chouman, M.: Hybrid metaheuristics for industry 5.0 multi-objective manufacturing and supply chain optimization. In: 2024 21st Learning and Technology Conference (L&T). IEEE, pp. 245–249 (2024)
7. Seydanlou, P., Jolai, F., Tavakkoli-Moghaddam, R., Fathollahi-Fard, A.M.: A multi-objective optimization framework for a sustainable closed-loop supply chain network in the olive industry: hybrid meta-heuristic algorithms. Expert Syst. Appl. **203**, 117566 (2022)
8. Momenitabar, M., Ebrahimi, Z.D., Ghasemi, P.: Designing a sustainable bioethanol supply chain network: a combination of machine learning and meta-heuristic algorithms. Ind. Crops Prod. **189**, 115848 (2022)
9. Mahmud, S., Abbasi, A., Chakrabortty, R.K., Ryan, M.J.: A self-adaptive hyper-heuristic based multi-objective optimisation approach for integrated supply chain scheduling problems. Knowl.-Based Syst. **251**, 109190 (2022)
10. Farmand, N., Zarei, H., Rasti-Barzoki, M.: Two meta-heuristic algorithms for optimizing a multi-objective supply chain scheduling problem in an identical parallel machines environment. Int. J. Ind. Eng. Comput. **12**(3), 249–272 (2021)
11. Faramarzi-Oghani, S., Dolati Neghabadi, P., Talbi, E.-G., Tavakkoli-Moghaddam, R.: Metaheuristics for sustainable supply chain management: a review. Int. J. Prod. Res. **61**(6), 1979–2009 (2023)
12. Coelho, D., Madureira, A., Pereira, I., Gonc, R., et al.: Multi-objective evolutionary algorithms and metaheuristics for feature selection: a review. Int. J. Comput. Inf. Syst. Ind. Manag. Appl. **14**, 12 (2022)
13. Amodeo, L., Prins, C., Sánchez, D.R.: Comparison of metaheuristic approaches for multiobjective simulation-based optimization in supply chain inventory management. In: Applications of Evolutionary Computing: EvoWorkshops 2009: EvoCOMNET, EvoENVIRONMENT, EvoFIN, EvoGAMES, EvoHOT, EvoIASP, EvoINTERACTION, EvoMUSART, EvoNUM, EvoSTOC, EvoTRANSLOG, Tübingen, Germany, April 15–17, 2009. Proceedings. Springer, pp. 798–807 (2009)
14. Canales-Bustos, L., Santibañez-González, E., Candia-Véjar, A.: A multi-objective optimization model for the design of an effective decarbonized supply chain in mining. Int. J. Prod. Econ. **193**, 449–464 (2017)
15. Elfarouk, O., Wong, K.Y., Wong, W.P.: Multi-objective optimization for multi-echelon, multiproduct, stochastic sustainable closed-loop supply chain. J. Ind. Prod. Eng. **39**(2), 109–127 (2022)
16. Gholian-Jouybari, F., Hashemi-Amiri, O., Mosallanezhad, B., Hajiaghaei-Keshteli, M.: Metaheuristic algorithms for a sustainable agri-food supply chain considering marketing practices under uncertainty. Expert Syst. Appl. **213**, 118880 (2023)
17. Ala, A., Goli, A., Mirjalili, S., Simic, V.: A fuzzy multi-objective optimization model for sustainable healthcare supply chain network design. Appl. Soft Comput. **150**, 111012 (2024)
18. Billal, M.M., Hossain, M.M.: Multi-objective optimization for multi-product multi-period four echelon supply chain problems under uncertainty. J. Optim. Ind. Eng. **13**(1), 1–17 (2020)
19. Abdi, A., Abdi, A., Fathollahi-Fard, A.M., Hajiaghaei-Keshteli, M.: A set of calibrated metaheuristics to address a closed-loop supply chain network design problem under uncertainty. Int. J. Syst. Sci. Oper. Logist. **8**(1), 23–40 (2021)
20. Avci, M.G., Selim, H.: A Multi-objective, simulation-based optimization framework for supply chains with premium freights. Expert Syst. Appl. **67**, 95–106 (2017)
21. Aslam, T.: Analysis of manufacturing supply chains using system dynamics and multi-objective optimization. Ph.D. thesis. University of Skövde (2013)

Identifying the Influential Factors for Mutual Fund Investments

Himani Grewal, Sapna Yadav, Avinash Saxena, and Meenakshi Tyagi

Abstract A mutual fund (MF) is generally called an arrangement under which the money from those having is collected for common investments objectives and thereafter invests their savings into different available assets as per the objective of the schemes. The investor can directly buy these assets from the market or with the help of financial intermediaries who are the experts of their field. Here the mutual funds work like a financial intermediary. A significant growth has been seen in the field of mutual fund industry in recent times. This study's primary goal is to conclude the quite a few essential elements that affect investors' decisions before investing in mutual funds in the Uttar Pradesh region of Moradabad. In the present study a cross-sectional survey research design was employed. The population includes various persons/consumers of financial instruments of Moradabad District of Uttar Pradesh out of which 130 respondents were selected by convenience sampling method. For data analysis SPSS 20 is used. Factor analysis is used to identify the various influential factors. It has been found that there are various factors which are considered by the investors before making an investment and with the help of factor analysis the researchers identified a list of six factors which are; benefits of investments, portfolio properties, future prospects of the investments, post purchase facilities, broker suggestions and fund ratings.

Keywords Mutual fund · Investor · Investor buying behaviour

H. Grewal (✉) · A. Saxena
Department of Management, Moradabad Institute of Technology, Moradabad, Uttar Pradesh, India
e-mail: drhimanigrewal@gmail.com

S. Yadav
Jaypee Business School, JIIT Noida, Noida, Uttar Pradesh, India

M. Tyagi
KIET School of Management, Muradnagar, Uttar Pradesh 201206, India

S. Kautish et al. (eds.), *Strategy Analytics for Business Resilience Theories and Practices*, Information Systems Engineering and Management 33,
https://doi.org/10.1007/978-3-031-82369-5_10

1 Introduction

Mutual fund, a financial vehicle, is an investment approach for those investors who congregate their savings for investing in to a variety of securities for example shares, debentures, bonds, capital and short term money market securities and other assets. The investors select the mutual fund with the intention of getting smart yields and enhancing the value. These are managed by qualified managers who distribute client funds among different assets in accordance with the scheme's investment goal. The optimal Risk Adjusted Return (RAR) is what a professional fund manager strives to deliver. Investors split the profit and loss according to the percentage at which they choose to allocate their investment. Mutual funds often offer a range of strategies catering to different investing goals. Mutual funds come in a variety of forms, including money market funds, equity, hybrid or debt funds, growth or income funds, liquid funds, bond funds, and open-end and closed-end funds. In order to maximize risk-adjusted returns and diversify risk, the investor looks for mutual funds as the most suited option. Generally speaking, a MF is an arrangement which aggregates the individual's savings with similar investing goals and allocates those funds to various accessible assets in accordance with the schemes' objectives. The investor has two options for purchasing these assets: directly from the market or with the assistance of financial intermediaries that possess in-depth knowledge and competence. In this case, mutual funds function as a kind of middleman. The average person can safely invest in mutual funds even if they lack the necessary background knowledge and experience in investing, risk and return related to assets, and other financial jargon. The four components of a mutual fund—the sponsor, trustee, asset management firm, and custodian—function as a trust. The sponsor forms a Board of Trustees in addition to creating the trust. For the advantage of the mutual funds holders, the trustees hold on to the fund's assets. The authorized corporate entity for money mobilization and securities investments is the asset management business. A custodian is an organization that is registered with the SEBI and is responsible for holding the securities of various mutual fund schemes.

All four components oversee the mutual fund. In order to offer the advantages of guaranteed returns, competent securities management, and risk diversification, mutual funds were introduced to India in 1963–1964. A multitude of factors influence the choice of the investor to decide what investments to make. The main goal of the research was to pinpoint the important variables that influence an investor's choice of assets. A wide range of criteria are involved, including as market expertise, risk, return, wealth generation, fund transparency and flexibility, and many more. The researchers made an effort to pinpoint the key elements influencing investors' choices.

2 Study Objectives

The primary objective for this study is to identify the various factors which are affecting the investor behavior while investing in mutual funds in Moradabad region of Uttar Pradesh.

3 Literature Review

Mishra [1] in his study titled as "Consumer demographics and the attitude towards mutual fund" identified the awareness and perception of investors towards MF and also identified influential factors to invest in MF. The researcher has taken sixteen variables for the study, tested factor analysis on the data set of 250 respondents, and concluded that there are four factors mainly the trust or belief, the risk, self-concept, and terminal value which are mainly considered by the investors before making investments in mutual funds.

Desigan et al. [2] studied on the opinion of female investor towards investing in mutual funds and identified that female investor have limited knowledge about investments, the process of investments, market situations, associated risk, investment assessments and grievance rederessal mechanism that is why they are mostly unsure and uncertain about investing their savings into mutual funds. Singh and Jha [3] in their study titled as "An empirical study on awareness and acceptability of mutual fund" identified that return, liquidity and safety of the funds are the main factors which are well thought-out by the investor previous to making investment in mutual funds and also they have identified that the investors have a limited knowledge about the SIP (Systematic investment plans). Bajracharya and Bilas [4], in their study the relationship between the demographic variables and socio economic variables has been found out and the investor attitude towards mutual fund investment. No association has been found between gender, age, income and educational level and attitude regarding MFIs. They have identified in their study that the sources of information like; annual reports, newspaper, magazine, broker, prospectus plays a very important role in decision making and broker advice is the most significant source.

Taparia and Babu [5] analyzed the attitude of the investors regarding the mutual fund investments and identified the role of demographic variable in the same. They have also examined that how the investment decision have influenced by some company specific factors, economic factors, choice of funds and investor risk taking ability.

They did their study on the data set of 120 respondents based in Kolkata and concluded that income level is very important factor in the risk taking capacity of the investors. They concluded that choice of fund is directly affected by the investors' buying capacity. Mutual fund is a safe mode of investment as per Kumar [6]. The researcher also identified investors have optimistic outlook regarding the mutual funds investments and also the past return performance of the fund has been properly analysed by the investor. Deo and Jagtap [7] evaluated the role of demographic factors on decision making of academicians on mutual fund investments. They have identified that unmarried earning individual does not any financial responsibility and they avoided huge investments on mutual fund but married individuals have invested a lot just to have another source of income. They have also found that male investors are more sensitive and aware as compared to female investors about the mutual fund investments. Singh [8] in his study identified the impact of demographic factors on investor attitude on a data set of 250 respondents. The researcher identified that gender, income and education level significantly influence the investor buying decision while age and occupation were not found to be influential. They have also identified that the return and liquidity of the fund were perceived more attractive benefits by the investors followed by flexibility of the fund, affordability of the investment and transparency.

4 Research Methodology

The primary focus of the current study is to identify various responsible factors influencing investor behavior regarding mutual fund investments in Moradabad region of Uttar Pradesh. Descriptive research design is used in current study followed by a quantitative research strategy for collecting the data and testing the hypotheses developed on the basis on literature reviewed.

Primary data has been used in the study which is collected by a questionnaire. The questionnaire contained two parts; part A and part B. Part A included the demographic details whereas the part B contained different influential factors adopted from the literature review which were considered by the investors before making any investment toward mutual funds. the second part of the questionnaire contained in total of 26 variables which were evaluated on a 5 point Likert scale ranging between strongly disagree to strongly agree. Cross sectional survey research design is used in the present study. The population includes various persons/consumers of financial instruments of Moradabad District of Uttar Pradesh out of which 130 respondents were selected by convenience sampling method. In order to analyze data and test the hypotheses, SPSS 20 was used to apply the relevant statistical techniques. Factor analysis has been used to extract the various influential factors.

5 Data Analysis and Findings

For every variable, the Cronbach Alpha α value was assessed independently. A Cronbach's alpha i.e., more than 0.6 is acceptable, indicating that the questionnaire's reliability is in a adequate range. For the total of 29 variables, the calculated Cronbach's α value is 0.912, indicating good data reliability for the study. From Table 1 descriptive statistics of the studied sample can be discussed. It is visible from Table 1, that 64% respondents included in the study were male and 36% were female, 60% respondents were married and 40% of the respondents were unmarried, 20% of the respondents fit in to 31–40 years age group, 64% respondents belonged to 31–40 years age group while only 16% respondents were from 41 to 50 years of age group. It has been found from the data that 29% respondents were from private job, 20% were professionals and 16% were from government job, 12% were housewife category, 10% of the respondents belonged to agriculture category. It has been found that 46% of the respondents were having intermediate qualification while 34% were post graduate. The data showed that 44% respondents having up to Rs. 2 lakh as family income (Tables 2 and 3).

Table 1 Descriptive statistics (n = 130)

Demographic Details of the participants		
Demographics		%
Gender	Male	64
	Female	36
Marital status	Married	60
	Unmarried	40
Age group	21–30	20
	31–40	64
	41–50	16
Designation	Business person	7
	Professional	20
	Private job	29
	Government job	16
	Student	6
	House wife	12
	Agriculture	10
Qualification	No formal education	10
	Intermediate	46
	Graduate	10
	Post graduate	34
Annual income	Up to Rs. 2 Lakh	44
	Rs. 2–5 Lakh	34
	Rs. 5–8 Lakh	22

Table 2 Results of KMO and Bartlett's test (Compiled by the authors)

KMO measure		0.854
Bartlett's test	Chi-Square	2014.765
	Sig	0.000

Table 3 Result of Factor loading of identified key factors (Compiled by the authors)

S. No.	Influential factors	Component					
		1	2	3	4	5	6
M4	Liquidity	0.821					
M3	Past dividend history	0.782					
M6	Wealth creation	0.747					
M2	Income security	0.743					
M5	Tax reduction	0.739					
M1	Fund safety	0.667					
M13	Transparency	0.606					
M21	Child marriage		0.928				
M23	Retirement		0.907				
M22	Purchase of new assets		0.884				
M20	Child education		0.771				
M19	Sustainable performance		0.631				
M25	Good portfolio			0.798			
M24	Fund size			0.779			
M26	Reputation of fund manager			0.744			
M8	Regular income			0.691			
M16	Risk taking ability			0.671			
M7	Risk protection			0.500			
M17	Current market				0.775		
M14	Grievance				0.688		
M18	Redemption facilities				0.682		
M10	Broker advice					0.802	
M9	Flexibility					0.761	
M12	Type of the fund					0.568	
M15	Good rating						0.785
M11	Past performance of the fund						0.688
	Total factor loading	5.105	4.120	4.184	2.145	2.132	1.40

6 Factor Analysis

The study has been made to identify the various factors which are influencing the investor behaviour towards MF investments in Moradabad region of Uttar Pradesh. There were 26 variables as influential factors under the study. The table no. 2 highlights the results given by KMO and Bartlett's Test of testing the sampling adequacy before conducting a factor analysis which is 0.854 which is higher than the required value 0.7 and shows that data can be used for factor analysis. The next table no 3

Table 4 Factors rank based on satisfaction

Factors	Factor loading	Rank
Benefits of investments	5.105	1
Future prospects	4.12	3
Portfolio properties	4.184	2
Post purchase facilities	2.145	4
Broker suggestion	2.131	5
Fund rating	1.473	6

shows the factors grouping or factor loading by Varimax method. By running factor analysis all the 26 factors has been grouped under six factors.

As we can see from the table that the first factor identified with the benefits of having investments like; liquidity, transparency, wealth creation, income security, tax deduction and fund safety, these all are grouped under factor 1 and labelled as "benefits of investments". The second factor has been identified the future prospects of doing any investment like; child education, child marriage, retirement planning, purchasing of any new assets and the sustainable performance of the fund and has been labelled as "Future Prospects". The third factor grouped all the factors related to portfolio properties like; fund size, good portfolio, reputation of fund manager, risk taking ability, risk protection and has been labelled as "Portfolio Properties". The fourth factor has been grouped together current market, grievance solving and redemption facilities which has been labelled as "Post Purchase Facilities". The fifth factor has been grouped roker advice, flexibility and type of the fund which has been labelled as "Broker Suggestion" and the last factor has been grouped good rating of the fund and past performance of the fund which has been labelled as "Fund Rating". Table 4, gives a clear picture of investors preferences of the various before investing and gives the ranking it shows that benefits of investing is ranked first by the respondents while fund rating is ranked last.

7 Conclusion

The current study provides a general idea about the factors which are considered by the individual investor before selecting any mutual fund. The objective of this research was to analyze and identifying major factors which are responsible for investor buying decision. It has been observed that there are various factors which are considered by the investors before making an investment and with the help of factor analysis the researchers identified a list of six factors which are again divided into different sub factors. As far as the influential factors of investor behaviour concerned the "Benefits of Investments" are highly considered by the investors followed by portfolio properties, future prospects of the investments, post purchase facilities, broker suggestions and fund ratings. Investors identify a number of factors and all these factors affect the decision of the investors.

References

1. Mishra, R.: Consumer demographics and the attitude towards mutual fund. J. Manage. Res. Anal. **5**(1), 42–47 (2020)
2. Desigan, C.G., Kalaiselvi, S., Anusuya, L.: Women investors' perception towards investment-An empirical study. Ind. J. Mark. **36**(4) (2006)
3. Singh, B.K., Jha, A.K.: An empirical study on awareness and acceptability of mutual fund. In: Regional Student's Conference, ICWAL, pp. 49–55 (2009)
4. Bajracharya, R.B.: A study of investors' attitude towards mutual fund in Kathmandu City, Nepal. J. Adv. Acad. Res. **4**(1), 1–11 (2017)
5. Taparia, S., Babu, R.C.: Impact of demographic factors on investment attitude of mutual fund investors in Kolkata. Int. J. Bus. Manage. Invention **8**(3), 11–18 (2019)
6. Kumar, D.R.: Investors attitude towards mutual funds. Int. Res. J. Manage. Commerce **04**(11), 2348–9766 (2017)
7. Deo, R., Jagtap, U.R.: Effect of demographic variables on mutual fund investment decision among academicians of Indore City. Int. J. Manage. Soc. Sci. **1**(1), 30–38 (2017)
8. Singh, B.K.: A study on investors' attitude towards mutual funds as an investment option. Int. J. Res. Manage. **2**(2), 61–70 (2012)

The Social Media Playbook: Strategies for Building Brand Loyalty and Engagement

S. Mahalakshmi, H. Bharath, and Sandeep Kautish

Abstract "The Social Media Playbook: Strategies for Building Brand Loyalty and Engagement" is a crucial manual for using social media to increase engagement and brand loyalty. The chapter highlights the need of customizing methods to the specific aspects of each major social media platform by examining its distinctive advantages. Examples of these strengths include Instagram's visual-centric approach, LinkedIn's efficacy in B2B marketing, and Instagram's captivating short-form videos. The chapter emphasizes how crucial it is to produce excellent, pertinent material and to schedule it well in order to preserve consistency. To evaluate the effectiveness of social media campaigns and make data-driven modifications, it provides information on key performance indicators (KPIs) such as engagement rates, reach, impressions, follower growth, and conversion rates. The chapter talks about how new technologies like augmented reality (AR), virtual reality (VR), and artificial intelligence (AI) will affect social media involvement in the future. AR and VR provide immersive experiences that can deepen consumer relationships with brands, while AI improves automation and personalization. Together with the expanding contribution of social commerce to the smooth operation of online stores, the evolution of influencer marketing toward more genuine, long-term collaborations is examined. Maintaining client trust is emphasized as requiring adherence to legislation such as the CCPA and GDPR, with a focus on data protection and security. The chapter also emphasizes how important user-generated content (UGC) is for creating a sense of community and repeat business. All things considered, this chapter offers practical advice and techniques for negotiating the ever-changing social media scene, adjusting

---

S. Mahalakshmi (✉)
Faculty of Management Studies, CMS Business School, Jain (Deemed-to-Be) University, Bengaluru, India
e-mail: mahalakshmi.s@cms.ac.in

H. Bharath
Marketing area, Faculty of Management Studies, CMS Business School, Jain (Deemed-to-Be) University, Bengaluru, India
e-mail: bharath.h@cms.ac.in

S. Kautish
AIT CSE, Chandigarh University, Mohali, Punjab, India

S. Kautish et al. (eds.), *Strategy Analytics for Business Resilience Theories and Practices*, Information Systems Engineering and Management 33,
https://doi.org/10.1007/978-3-031-82369-5_11

to new trends, and successfully interacting with consumers to foster enduring brand loyalty.

Keywords Social media · Brand loyalty · Engagement strategies · Key performance indicators (KPIs) · Influencer marketing · Augmented reality (AR) · Virtual reality (VR) · Data privacy

1 Introduction

Social media has completely changed modern brand strategy and is now necessary for interacting with consumers, fostering brand loyalty, and increasing engagement. Social media, in contrast to traditional marketing, provides dynamic, real-time contact that enables firms to establish a personal connection with customers [25] Social media sites such as Facebook, Instagram, Twitter, LinkedIn, and TikTok offer distinctive chances to target particular audiences with customized content. Customers expect businesses to have a strong social media presence in the modern digital world. It promotes community and trust by enabling brands to highlight their services and values. Credibility and visibility are increased through influencer collaborations, interactive campaigns, and user-generated content. Through analytics, social media also provides useful information that helps firms fine-tune their plans in response to real-time input [24].

The transition of traditional marketing to digital platforms signifies a momentous change in the way firms interact with their clientele. Print, radio, television, and outdoor advertising were all part of traditional marketing, which mostly provided one-way communication with little engagement. Although it was successful in reaching a large audience, it was unable to offer individualized experiences and engage users in real time [23]. On the other hand, because digital platforms allow for interactive interaction and two-way communication, marketing has completely changed. Through the use of data analytics and social media, search engines, email marketing, and websites, organizations can precisely target particular audiences and gain insight into their interests and behavior. As a result of this change, marketing is now more quantifiable and flexible, enabling real-time performance tracking and strategy modification. Digital platforms have also made marketing more accessible and affordable, giving small businesses the edge over bigger players. Digital content's immediacy and virality facilitate the quick spread of information, while influencer relationships and user-generated content increase reach and trustworthiness. Digital marketing has, all things considered, resulted in more dynamic, data-driven, and consumer-focused tactics, radically altering the way firms interact with their target audiences and develop relationships with them [3].

This chapter seeks to accomplish three goals: it will clarify the critical role that social media plays in modern brand strategy; it will identify and evaluate the advantages and target audiences of the major social media platforms; and it will offer suggestions for creating content strategies that work. Readers will be able to appreciate how social media can improve engagement and brand loyalty by knowing its importance [23]. The chapter will examine the main social media platforms and provide readers with information about their distinct characteristics and user bases. This information can then be used to inform targeted marketing campaigns. The chapter will also help readers create captivating content that appeals to their target audience and builds stronger bonds and brand loyalty [3].

Readers will leave this book chapter with a thorough grasp of the strategic role that social media plays in establishing and preserving brand loyalty. The readers will acquire useful information regarding the unique benefits of different social media sites and the best ways to use them. Readers may improve their brand's social media presence and engagement by learning how to create engaging content and encourage real conversations [25]. With the help of the chapter's insights on analytics, community development, and personalization, readers will be able to develop dynamic, data-driven strategies that adjust in real-time to trends and feedback. In the end, this chapter will provide readers the confidence to confidently traverse the digital terrain, ensuring that their business stays competitive and relevant [24].

2 Understanding Brand Loyalty in the Digital Age

The dynamic interaction between businesses and consumers on social media platforms has shaped the complex concept of brand loyalty in the digital age. In contrast to conventional brand loyalty, which is based on consumer happiness and constant product quality, digital brand loyalty is based on community creation, interaction, and personalization. Through personalized experiences and meaningful interactions, social media has emerged as an essential tool for brands looking to cultivate customer loyalty. Brands may foster a feeling of community and confidence among their target audience by utilizing influencer relationships, user-generated content, and interactive marketing. Through rapid resolution of issues and feedback, real-time communication, and responsiveness on social media improve consumer happiness and loyalty [9].

The utilization of digital techniques can result in strong brand loyalty, as evidenced by case studies of prosperous firms. To build a devoted following of customers, companies such as Starbucks and Nike, for example, have become experts in producing interesting and compelling content for social media. In the digital era, it is imperative to have the capacity to create dynamic virtual communities where customers feel appreciated and involved. Through social media, brands that can foster and maintain brand loyalty are better positioned to prosper as customers seek out more genuine and engaging experiences. To flourish in the contemporary economy, organizations must ultimately comprehend and capitalize on the power of digital brand loyalty [4]

2.1 Definition and Components of Brand Loyalty

When a consumer consistently chooses and again buys a specific brand's goods or services over time, even in the face of competing options, this is referred to as brand loyalty. Because of their deep emotional bond, satisfying prior experiences, and sense of worth, devoted customers are less susceptible to price adjustments and aggressive marketing campaigns. Emotional connection, customer happiness, and brand advocacy are the main drivers of brand loyalty, and they all have a significant impact on the development of enduring relationships between consumers and businesses. A consumer and a brand can form an emotional relationship through relatable brand values, positive experiences, and compelling storytelling. Customers are more likely to stick with a brand they are attached to because it builds trust and a sense of community. Customers who sense a personal connection to a brand's identity and mission can become strong and devoted supporters of brands that emotionally connect with their audience [15].

Customer happiness is yet another essential element of brand loyalty. Meeting or surpassing customer expectations regularly promotes loyalty and repeat business. Customers are more inclined to stick with a brand over competitors when they are pleased with the goods or services that it offers. Long-term success depends on a positive reputation and higher customer retention rates, both of which can be attained through high customer satisfaction levels. When happy and devoted consumers actively spread the information about a brand to others, usually via social media or word-of-mouth, this is known as brand advocacy. Because suggestions from reliable sources have greater credibility and influence than those from the brand itself, this advocacy is particularly potent. Brands that have devoted supporters can reach a wider audience and draw in new clients, which improves their standing in the marketplace [10].

For a brand to succeed over the long haul, brand loyalty is an intricate and varied phenomenon. Businesses may create a devoted client base that underpins long-term growth and competitive advantage by building emotional attachment, guaranteeing high customer happiness, and stimulating brand advocacy. A stronger, more resilient brand in today's cutthroat industry might result from an Understanding of and attention to these elements.

2.2 The Role of Social Media in Shaping Consumer Perceptions and Loyalty

Social media is essential in today's digital environment for influencing consumer views and building brand loyalty. With its help, brands can use it as a platform to communicate meaningfully with their audience in real-time and develop a community around their offerings. Social media's accessibility and immediacy enable brands to speak to customers directly about their beliefs, offers, and stories, changing their attitudes and views in the process. Another important element that shapes loyalty and opinions is engagement. Social media facilitates bidirectional communication, allowing firms to engage with customers face-to-face. This communication can happen in several ways, such answering questions from customers, joining in on debates, and replying to comments. Brands may increase client loyalty by exhibiting their dedication to customer satisfaction through responsiveness and attentiveness. Influencer collaborations are also quite important. The brand's credibility and reach can be increased by working with influencers who share the brand's values and target market. Influencers frequently have devoted fan bases, and their recommendations have a big impact on how consumers perceive brands and foster brand loyalty [27].

Another effective strategy on social media is user-generated content, or UGC. A level of authenticity and trust that is not possible with traditional advertising is added when customers share their experiences with a brand. As social proof, user-generated content (UGC) encourages people to interact and trust a company. Social media's interactive quality and extensive reach render it a vital instrument for molding customer attitudes and fostering brand allegiance. Brands may leave a good, lasting impression that promotes loyalty and propels long-term success by utilizing content, engagement, influencer relationships, and user-generated content [13].

2.3 Real-Time Instances of Brands that Have Successfully Built Loyalty Through Social Media

Numerous organizations have effectively established brand loyalty via social media by utilizing diverse tactics to captivate their audience, generate significant exchanges, and cultivate a sense of community [13]. Here are a few examples from real-time:

- **Nike**

One of the best examples of a company using social media to foster customer loyalty is Nike. Nike regularly produces motivational and inspirational content that connects with its audience through its "Just Do It" campaign and other efforts. By encouraging users to share their accomplishments and stories, Nike builds a community of devoted fans by utilizing athlete sponsorships and user-generated content. The company's involvement in social concerns and inclusive marketing initiatives deepens its emotional bond with customers.

- **Starbucks**

Starbucks has done a fantastic job of leveraging social media to increase client loyalty. Customers feel appreciated and heard when a brand responds to their questions and comments on social media sites like Instagram and Twitter. Starbucks also announces the release of new products, offers exclusive discounts, and promotes its rewards program on social media. Customers submit images of their Starbucks red cups as part of the “Red Cup” promotion, which encourages community building and produces a lot of user involvement throughout the Christmas season.

2.4 GoPro

GoPro has used user-generated material to build a devoted following. The company invites users to use the hashtag #GoPro when posting images and videos taken using GoPro cameras on social media. This not only demonstrates the product’s potential but also fosters the growth of an enthusiastic community of adventure seekers around the brand. GoPro gives its customers a sense of recognition and community by regularly showcasing user material on its social media channels.

3 Key Social Media Platforms and Their Unique Strengths

- **Instagram**: Well-known for its visually stimulating material, Instagram works particularly well at capturing the attention of younger consumers, especially millennials and Generation Z. Its features—like Stories, Reels, and IGTV—make it possible to deliver content dynamically and imaginatively. Instagram is a potent medium for firms that focus on aesthetics, lifestyle, fashion, and travel because of its visual emphasis. Seamless e-commerce experiences are also made possible by the integration of shopping elements [5].
- **Facebook**: With a user base spanning multiple age groups, Facebook is among the most adaptable and oldest platforms. Its power resides in creating communities via pages and groups, providing a forum for people with similar interests to interact. The platform is perfect for focused marketing initiatives because of its extensive analytics and advertising features. Facebook’s usefulness for businesses is further increased by its connection with other apps and services [5].
- **Twitter**: The preferred medium for news, trends, and public conversation is Twitter due to its real-time communication features. Users can exchange and digest information rapidly thanks to its succinctness and immediacy. Public relations, customer service, and conversational interaction are the areas where Twitter excels. Reach and community involvement are increased through the use of hashtags and popular themes [5].
- **LinkedIn**: Being the leading platform for professional networking, LinkedIn is great at bringing together professionals from the same industry, developing business-to-business connections, and exchanging content relevant to the industry. For companies aiming to reach professionals and decision-makers, it is crucial because of its power in career development and professional branding. Long-form content and thought leadership are also possible with LinkedIn’s publishing platform [5].

- **YouTube**: The most popular site for sharing videos, YouTube excels in long-form video content. For entertainment, vlogs, tutorials, and instructional content, it is a vital platform. There are plenty of options for discoverability and revenue thanks to YouTube's search engine capabilities and Google Ads integration [5].

"Key Social Media Platforms and Their Unique Strengths," a section of the chapter, will give readers the strategic knowledge they need to decide how best to use social media. Readers may more effectively match their social media strategies with their target demographics and specialized goals by knowing the unique characteristics and audience preferences of platforms such as YouTube, Facebook, Instagram, LinkedIn, TikTok, and Twitter [7]. With this knowledge, they will be able to efficiently customize their content so that it resonates with each platform's distinct user base and increases engagement. This section will also assist readers in making the most use of their resources by pointing out which platforms provide the highest ROI for various content kinds and marketing goals. This thorough understanding will ultimately enable readers to craft social media strategies that are more effective and impactful, improving the outcomes of their projects or companies [7].

3.1 Platform-Specific Strategies for Engagement and Loyalty

- **Instagram**: Create visually appealing and genuine material to increase interaction. For instance, National Geographic (@natgeo) engages its audience with compelling narratives and high-quality photographs. They regularly hold interactive Q&A sessions, publish user-generated content, and share behind-the-scenes photos. Reels can be used to display brief, entertaining videos, and Instagram Stories can be used to offer daily updates. Work together with influencers who share your ideals to gain credibility and access to their fan base [19].
- **Facebook**: Starbucks makes good use of Facebook by keeping an active page with frequent updates, interesting content, and community-building activities. To interact with their audience in real time, they regularly hold Facebook Live events. Create Facebook groups based on common interests as a way to promote a feeling of community. To reach particular demographics and retarget visitors who have expressed interest in your material, employ tailored adverts. To increase loyalty, swiftly answer messages and comments [19].
- **Twitter**: With its capacity to communicate in real time, Twitter is a great tool for dynamic conversations and timely updates with audiences. The Mumbai Police (@MumbaiPolice) is a well-known Indian example of creative Twitter engagement. Twitter has been a useful tool for the Mumbai Police to interact with the public, disseminate vital information, and forge positive public perception. They have a reputation for sending clever and lighthearted tweets that both educate and amuse their followers. During the COVID-19 pandemic, for example, Mumbai Police emphasized the significance of adhering to safety protocols by using witty

wordplay and analogies to popular culture. Oftentimes, their tweets become viral, greatly expanding their audience and involvement [19].

- **YouTube**: High-quality, educational, and entertaining material that is catered to the interests of your audience is the key to maximizing engagement and loyalty on YouTube. BB Ki Vines, a well-liked channel run by Bhuvan Bam, is a great example in viewer engagement. With his amusing and relevant sketches, Bhuvan Bam has gained a devoted following from his audience. On BB Ki Vines, a YouTube channel, humorous sketches about daily life are posted. Bhuvan Bam plays various characters in many of these sketches. He has established a close relationship with his audience because of his incredibly accessible content, which captures the subtleties of Indian family and societal interactions [19].

3.2 Demographic Insights and User Behavior Analysis

Developing successful social media campaigns requires a comprehensive knowledge of the characteristics and actions of your target audience. User behavior analysis looks at how users interact with material, including their preferences, engagement patterns, and activity times. Demographic insights include age, gender, geography, and hobbies [17].

Demographic Insights

- **Gender and Age**: Online activities and preferences differ throughout genders and age groups. In contrast to Facebook, which has a wider age range that includes older demographics, Instagram and TikTok mostly draw younger users, such as teens and young adults. People in their 25–49 age range choose LinkedIn.
- **Location**: Content can be better tailored to local interests and cultural quirks with the use of geographic data. For instance, realizing that a sizable percentage of your YouTube subscribers are from India may inspire you to produce material that speaks to Indian trends and culture.
- **Interests and Hobbies**: Knowing your audience's interests and pastimes enables you to provide more individualized content. Targeted marketing efforts are made possible by the comprehensive user interest data that social media platforms like Facebook and Instagram offer.

User Behavior Analysis

- **Engagement Patterns**: By examining how users engage with your material—through likes, shares, comments, and views—you may determine what kind of content your audience finds most compelling. YouTube statistics, for example, can display the most popular topics or formats among the videos, as well as which ones have greater involvement [14].
- **Activity Times**: Posting schedule optimization for optimum reach and engagement depends on knowing when your audience is most active. Posting at the

right moments to reach your audience online is made possible by Instagram and Facebook's insights on peak activity periods [14].
- **Content Preferences**: By examining which types of content perform best, such as videos, images, or articles, you can tailor your content strategy accordingly. For example, if your LinkedIn audience prefers in-depth articles, focusing on publishing detailed posts and thought leadership pieces can drive better engagement [14].

4 Content Strategies for Engagement

Building a devoted following and fostering deep connections on social media platforms need producing interesting material. Finding out about your audience's demographics, hobbies, and internet habits is essential first. This makes it possible to provide them with content that is more specifically tailored to their interests. Engagement is greatly influenced by visual content, and attention can be successfully drawn in by employing GIFs, infographics, videos, and high-quality photos. Using these images to illustrate a story can also elicit feelings from viewers, which humanizes your brand.

Engagement is further increased by utilizing platform-specific features. To enhance follower engagement, use interactive stickers such as polls and Q&A sessions. You can also share dynamic content on Instagram by using Reels, Stories, and IGTV. Facebook's real-time engagement and community-building potential are derived from group interactions and live streaming. Whereas LinkedIn benefits from professional content and group engagement, Twitter concentrates on current conversations and hashtag use. While YouTube's long-form, educational material can cultivate a devoted following through playlists and series, amusing videos and trending challenges encourage discoverability and user engagement.

Another effective tactic is to support user-generated material. Community and social proof can be promoted through holding challenges, contests, and showcasing content from followers. In order to keep your audience interested and looking forward to fresh information, you should stick to a regular posting schedule that is adjusted according to the peak activity periods on each platform. By involving your audience in real time, interactive content such as surveys, polls, and live streaming can further improve engagement. Lastly, you can make sure that your content is engaging and relevant by routinely analyzing analytics to understand content performance and modifying your strategy in response to platform improvements and new trends.

4.1 Importance of High-Quality, Relevant Content

In the context of audience engagement and digital marketing, high-quality, pertinent content is essential and provides a number of vital advantages. Above all, it is aesthetically pleasing, educational, and worthwhile, which draws in and holds the interest of the viewer. By receiving likes, shares, comments, and following, this kind of material promotes user interaction. When compared to a less polished video, a well-produced YouTube video instructional with crystal-clear images and useful information is likely to enhance audience retention and subscriptions [23].

Furthermore, producing top-notch content increases credibility and trust among your audience. Accurate and helpful information positions your brand as a reliable authority in your industry. A thorough and well-researched blog article on investing methods by a financial advisor, for instance, can build confidence and establish the brand as a trustworthy authority. Additionally, search engine optimization (SEO) is greatly aided by high-quality content. Search engine rankings are improved by content that incorporates pertinent keywords and keeps readability high. This increases the visibility of your website and generates natural traffic. An informative, well-written travel blog has a higher chance of ranking for inquiries about travel [26]. Creating content that is specifically tailored to the needs and interests of your audience encourages more engagement and deeper connections. You may produce engaging content by targeting their hobbies or pain concerns. A fitness company that provides customized training regimens, for instance, can accommodate a range of fitness requirements and levels. Superior content also sets your brand apart from rivals by giving it a distinct voice and personality. Red Bull, for example, distinguishes itself from other companies and maintains its adventurous image with its extreme sports videos [16].

Lastly, conversions and lead generation can be boosted with pertinent content. Offers of useful materials like eBooks, whitepapers, or product demonstrations might persuade visitors to perform desired activities like subscribing to newsletters or getting a product. To summarise, the long-term advantages of investing in relevant, high-quality content include increased trust, driving traffic, engagement, and conversions [12].

4.2 Types of Content that Drive Engagement

To increase engagement, you must provide material that speaks to your audience. With their own merits, several kinds of material can draw in viewers and stimulate conversation.

- **Videos**: Visual media is very adaptable and captivating. Dynamic visuals and succinct messages draw viewers in to short-form videos, like those seen on TikTok or Instagram Reels. Prolonged films, such as webinars or YouTube lessons, offer comprehensive knowledge and encourage closer relationships. A skillfully composed how-to video or a behind-the-scenes peek at your business, for example, can draw in visitors and encourage meaningful conversation [20].
- **Infographics:** Infographics condense difficult information into aesthetically pleasing, understandable visuals. They can convey important information or

insights quickly and are very shareable. An infographic with industry statistics or step-by-step instructions, for instance, can get shares and comments, which makes it an effective tool for increasing interaction [20].

- **Blog Posts**: Whether they offer insightful analysis, address frequently asked queries, or provide answers to issues, well-written and informative blog posts can significantly increase reader engagement. Adding interactive material, films, and other multimedia components can increase user engagement even further. One blog post that can spark conversations and shares is a comprehensive guide on digital marketing tactics that includes case studies and useful advice.
- **User-Generated material**: You may greatly increase engagement by encouraging your audience to produce and distribute material that is relevant to your brand. Social media posts, reviews, and testimonies are examples of user-generated content that fosters community and boosts authenticity. One way to increase engagement and create a feeling of community with your company is to hold a photo competition and display user entries on your social media pages [6].
- **Live Streaming**: Having conversations in real time via live streaming can give an impression of immediacy and urgency. Events, product demos, and live Q&A sessions provide you the chance to interact directly with your audience and build a stronger personal bond. An industry trends conversation or a live stream exhibiting a new product introduction, for example, can generate a lot of interest.

4.3 Creating a Content Scheduling and Maintaining Consistency

The key to effective digital marketing and audience engagement is developing a plan for content scheduling and sticking to it. Content generation may be streamlined with the aid of a well-structured content calendar, which guarantees timely, relevant, and strategy-aligned posts. Begin by listing important dates, including industry gatherings, product launches, and holidays, and then organize material around these occasions. For scheduling and organization, use apps like Trello, Google Calendar, or specialized social media management software like Hootsuite or Buffer. To keep your audience interested and waiting for your updates, you need to maintain a consistent flow of information, which is made possible by this preparation [26].

Posting consistently is crucial for establishing and preserving audience expectations and trust. Whether your posting schedule is daily, weekly, or bi-weekly, sticking to a regular schedule helps your audience become used to it and helps them anticipate new information. Stronger bonds and more regular interactions are fostered by this dependability. SEO benefits from consistency as well because search engines like material that is updated frequently, which raises your profile and ranking [20].

Apart from planning, make sure that the caliber of your content consistently meets the tone and principles of your brand. Providing value with every article is as important as maintaining a regular posting schedule. In order to keep your plan relevant and effective, keep an eye on engagement metrics and audience comments. Making adjustments to unanticipated developments or emerging trends without upsetting your overarching plan is made possible by using a content calendar. You can develop a devoted following, increase engagement, and have long-term success with your digital marketing initiatives by sticking to a well-planned schedule and producing high-quality content [6].

5 Measuring Success: Metrics and KPIs

Monitoring important metrics and KPIs that demonstrate the efficacy of tactics meant to foster brand loyalty and engagement is a necessary part of measuring success on social media. A crucial statistic is the engagement rate, which quantifies the amount of interaction—likes, comments, shares, and clicks—that your content receives in relation to the size of your audience. A high amount of engagement suggests that your material is connecting with readers and generating conversations. Additionally, significant are reach and impressions, which measure how many distinct individuals see your content overall and impressions, which count the number of times it is presented without any interaction [18]. These indicators assist in determining the effectiveness and visibility of the distribution of your material. Another useful KPI is follower growth, which indicates the gradual expansion of your audience base. Consistent increase in followers indicates that your content strategy is reaching new consumers and raising awareness of your business. Insights into how your audience views your brand can also be gained by examining consumer sentiment, which is the general tone of interactions and feedback. Sentiment analysis tools can tell you if a sentiment is good, negative, or neutral, which will allow you to modify your interaction and content strategy appropriately [22].

Conversion rate is a metric that indicates how well your social media efforts convert into real results by counting the proportion of people that complete desired actions, such buying something or subscribing to a newsletter. In comparison to competitors, share of voice measures the proportion of discourse about your industry or niche that mentions your brand, providing insight into the prominence and market position of your brand. Lastly, monitoring your client retention rate gives your insight into how successfully your tactics are encouraging continuous interaction and long-term loyalty. You may obtain important insights into the efficacy of your social media strategy by routinely analyzing these metrics and KPIs. This will help you make well-informed decisions that will increase engagement and strengthen brand loyalty [22].

5.1 Key Performance Indicators for Social Media Success

Key performance indicators (KPIs) are crucial measurements that assist in determining how well your plans are working toward your objectives on social media. The degree of interaction your material receives in relation to the size of your audience, including likes, comments, shares, and clicks, is measured by the engagement rate, which is a main KPI. A high percentage of involvement suggests that your material strikes a chord with readers and fosters deep conversations. Reach and impressions are also quite important; reach is the total number of distinct users who view your content, and impressions are the total number of times your material is shown without any kind of interaction. These metrics evaluate the effectiveness and visibility of the way your material is distributed. A consistent increase in followers indicates that your content strategy is reaching new audiences and raising brand awareness. Follower growth measures the evolution of your audience base over time [11]. Positive, negative, or neutral sentiment is revealed through customer sentiment research, which also helps you make changes to your engagement and content strategy. It also gives you insights into the general tone of interactions and feedback. By calculating the proportion of users who complete desired actions, such purchasing a product or subscribing to a newsletter, conversion rate indicates how well your social media efforts result in measurable results. Comparing your brand to competitors in a certain industry or niche, share of voice measures how much of the conversation revolves around your brand, showing its significance and position in the market. Lastly, customer retention rate measures how successfully your tactics encourage continuous involvement and long-term commitment. You can make well-informed decisions to improve brand loyalty and increase engagement by routinely monitoring these KPIs, which will provide you with insightful information about how well your social media initiatives are working [11].

5.2 Tools and Techniques for Tracking and Analysing Social Media Metrics

Combining specialized tools and methodologies is necessary for effective tracking and analysis of social media metrics in order to collect information, offer insights, and direct tactical adjustments. You can schedule posts, manage numerous accounts, and monitor interaction metrics with social media management tools like Hootsuite and Buffer [11]. These tools also provide detailed reports on follower growth, engagement, and reach. Each platform-specific native analytics tool, such Facebook Insights, Instagram Insights, Twitter Analytics, and LinkedIn Analytics, offers comprehensive information on post-performance, audience demographics, and engagement. Google Analytics analyzes user behavior, conversion rates, and overall campaign effectiveness to provide more comprehensive insights by tracking visits to your website via social media. Social listening solutions such as Mention and Brandwatch track brand mentions online and in social media, evaluating sentiment and seeing patterns instantly [21]. In-depth data are available through third-party analytics solutions like Sprout Social and Klear, which track interaction, social media trends, and influencer marketing effectiveness. Sentiment analysis technologies, such as Lexalytics and

Brand24, measure public opinion of your brand through natural language processing and offer insights into audience engagement and sentiment. The effectiveness of traffic generated by social media can also be understood by using heatmaps and click tracking tools like Hotjar, which show how users interact with your website. You can track and analyze social media metrics more efficiently by using these tools and strategies, and you'll get insightful data that will help you improve engagement, develop brand loyalty, and succeed on social media in general [11].

5.3 Adjusting Strategies Based on Performance Data

The effectiveness of traffic generated by social media can also be understood by using heatmaps and click tracking tools like Hotjar, which show how users interact with your website. You can track and analyze social media metrics more efficiently by using these tools and strategies, and you'll get insightful data that will help you improve engagement, develop brand loyalty, and succeed on social media in general. You can devote more resources to producing content of that type when you observe that particular formats, such infographics or videos, regularly result in greater interaction rates. In your content calendar, you should include more Instagram Stories if, for instance, they get more engagement than static posts [21]. In contrast, examine why a specific kind of content might not be connecting with your audience if it isn't working well and think about changing it up or substituting it with another format. Sentiment analysis and audience feedback are essential tools for plan modification. It can be useful to keep an eye on reviews, messages, and comments to spot recurring themes or issues. It's critical to resolve any unfavorable attitude that is identified in relation to a particular topic or campaign in a timely and open manner. Expansion or replication of successful aspects might be highlighted by positive feedback [28].

The timing of your posts is another important consideration. You can modify your posting schedule by looking at the times when your audience is most engaged. Plan additional content for the evening to optimize visibility and interaction if data indicates that engagement surges then.

Furthermore, A/B testing might aid in the improvement of your tactics. You can find out which versions of a post perform better by testing variations in content, structure, or timing. Then, you can use those learnings throughout your approach [28].

Finally, monitoring the performance of rivals can provide insightful information. Examining the strategies that your rivals employ can help you develop new ones or identify weaknesses in your strategy. By regularly modifying your social media strategy in response to performance data, you can make sure that your efforts continue to be efficient, pertinent, and in line with the preferences of your audience, which will eventually improve engagement and foster brand loyalty [8].

6 Future Trends in Social Media and Brand Loyalty

The social media and brand loyalty landscapes are expected to change significantly as technology advances. For companies looking to preserve and improve their online presence, keeping up with current trends is essential. Social media marketing is increasingly reliant on automation and artificial intelligence (AI). Large volumes of data can be analyzed by AI-driven systems to provide insights into customer behavior, forecast trends, and tailor content. With their ability to respond instantly and improve user experience, chatbots and automated customer support solutions will only get better. AI, for instance, can personalize responses by analyzing previous interactions, improving the effectiveness and personalization of customer service. Social media is about to undergo a transformation thanks to the immersive experiences that augmented reality (AR) and virtual reality (VR) technology offer. Companies can utilize augmented reality (AR) to provide virtual try-ons for clothing and cosmetics, giving consumers a preview of the items before they buy. Virtual reality (VR) may produce captivating virtual worlds for conferences, product debuts, and interactive narratives. Businesses may boost brand loyalty and set themselves apart from rivals by utilizing these technologies to create distinctive and captivating experiences [1].

Influencer marketing will keep expanding, but it will start to focus more on genuine, long-term relationships. The value of micro-influencers, who have tiny but very engaged audiences, will rise. Real engagement and trust can be sparked by their sincere relationships with their fans. To ensure more meaningful and reliable endorsements, brands should concentrate on developing long-term relationships with influencers who share their values and appeal to their target market. As social media and e-commerce become more deeply integrated, social commerce will grow significantly. In-app checkouts, shoppable posts, and live-stream shopping are examples of features that will proliferate and boost user pleasure by promoting conversions. Companies should employ analytics to monitor the success of their social commerce initiatives and optimize their social media profiles for shopping [9].

Businesses and social media platforms must emphasize data security and transparency as worries about data privacy and security continue to rise. It will be essential to abide by laws like the CCPA and GDPR. Brand loyalty will increase when trust is established through open data policies. Organizations must to maintain awareness of legislative modifications and put strong data security protocols in place. UGC (user-generated content) will continue to be a potent tactic for fostering a sense of community and trust. Companies ought to design campaigns that encourage consumers to share their stories. In addition to offering real information, user-generated content (UGC) encourages community and loyalty. Businesses should take a proactive stance, investing in new technology, learning from industry advancements, and continuously adapting to prepare for these future trends. Developing scalable and adaptable marketing [1].

7 Challenges and Solutions in Social Media Strategies

There are a number of obstacles to overcome when it comes to social media management, but using smart solutions can greatly improve your results. Creating high-quality material consistently is a major undertaking that requires imagination, effort, and money. Creating a content calendar that allows posts to be scheduled ahead of time aids in maintaining a consistent flow of information. This process can be streamlined with the use of tools like Hootsuite for scheduling and Canva for visual design [2]. Working with a varied team or contracting out to independent contractors can help reduce constraints associated with content development and introduce new views. Growing and involving your audience in the face of the ever-increasing volume of content on social media platforms presents another difficulty [9]. Focus on producing interesting, pertinent material that is suited to the tastes and interests of your audience to solve this. To find out what your audience finds engaging, regularly review engagement numbers and make any adjustments to your plan. Moreover, the visibility of your material may be affected by the frequent changes in social media algorithm. It is imperative to be informed about platform modifications and modify your approach accordingly. React swiftly when needed by tracking audience mood and trends in real-time with social listening tools like Brandwatch and Mention [28]. Another major problem is handling unfavorable reviews while preserving a favorable brand image. Potential problems can be transformed into opportunities to show your brand's dedication to client satisfaction by responding to unfavorable remarks fast and openly. Having strong crisis management procedures in place guarantees that you are equipped to deal with unanticipated bad press after it occurs. And last, because engagement and conversions have multiple facets, calculating the return on investment from social media initiatives can be challenging [21]. Employ all-inclusive analytics programs such as Sprout Social and Google Analytics to monitor specific metrics and evaluate the effectiveness of your social media marketing initiatives. Consistently analyzing these indicators aids in improving performance and strategy. You may improve your social media management and promote greater engagement and brand loyalty by applying strategic solutions to these problems [23].

8 Conclusion

Through our exploration of the complex world of social media marketing in "The Social Media Playbook: Strategies for Building Brand Loyalty and Engagement," we have brought attention to the tactics and resources that are crucial for cultivating meaningful customer engagement and strong brand loyalty. The main social media sites, their distinct advantages, and the kinds of content that encourage interaction have all been thoroughly covered in this chapter. We've talked a lot about the value of relevant, high-quality content and the necessity of consistency in content scheduling. The metrics and KPIs required to assess social media success and modify plans in response to performance information were also covered. Social media will undoubtedly be the most popular channel for interacting with customers in the future. Technology advancements like virtual reality, augmented reality, and artificial intelligence will completely change the way that organizations communicate with their customers. Whereas AR and VR will provide immersive experiences that help improve product visualization and storytelling, AI will provide more efficient and personalized customer care through chatbots and predictive analytics. Beyond drawing in and interacting with consumers, these technologies will also help brands build stronger emotional bonds with their target audiences. Customer purchases made directly through social media platforms will grow easier as influencer marketing transitions to more genuine, long-term collaborations and social commerce becomes more smooth. Moreover, companies must give transparency and data security top priority in order to gain and keep the trust of their target audiences as worries about data privacy grow.

This chapter has discussed how social media will continue to be the most important channel for customer engagement due to its wide reach and capacity to produce individualized and interactive experiences. Through the integration of cutting-edge technologies and adaptability to evolving market trends, brands can fully leverage social media to establish long-lasting connections with their clientele. Businesses will have the tools necessary to successfully navigate the always changing digital landscape and maintain their social media presence with the techniques and insights offered here.

9 Key Takeaways

- Recognizing Platform Strengths: Every social media platform has certain advantages that are tailored to particular facets of brand interaction. As an illustration, TikTok is best suited for short, interesting videos, Instagram is great for visual material, and LinkedIn is good for B2B networking. In order to maximize interaction, material must be tailored to the strengths of each platform.
- Excellent Content Is Essential: The key to increasing engagement is producing and preserving relevant, high-quality material. Content should speak to the needs and interests of your target audience while also being consistent with the voice of your brand. Maintaining a dependable brand presence is facilitated by content schedule consistency.
- Relevance of Metrics and KPIs: It's critical to gauge success using key performance indicators (KPIs) including engagement rates, impressions, reach, follower growth, and conversion rates. These metrics assist in making data-driven decisions by revealing how effective your social media initiatives are.
- Leveraging Innovative Technologies: Technological developments will influence social media patterns in the future. Social commerce will expedite the purchasing process, augmented reality (AR) and virtual reality (VR) will provide immersive

experiences, and artificial intelligence (AI) will improve automation and personalization. Using these technologies can increase brand loyalty and foster deeper levels of interaction.

- Changes in Influencer Marketing: Real, lasting connections are becoming more prevalent in influencer marketing. There's a growing demand for micro-influencers with large viewerships. Collaborating with influencers that genuinely connect with their audience and have similar values should be the primary goal for brands.
- Handling Data Privacy and Security: As worries about data privacy become more prevalent, companies need to give openness and strong data protection procedures top priority. Sustaining confidence and protecting consumer data requires adherence to laws like the CCPA and GDPR.
- UGC's (user-generated content): User-generated content can strengthen community and brand trust when it is promoted and used effectively. Customers are encouraged to feel involved and loyal via UGC, which offers genuine endorsements.
- Adaptability is Key: The social media world is ever-changing; therefore, adaptability is essential. It is essential to stay up to date on audience preferences, technological advancements, and trends in order to adjust tactics and keep a competitive advantage.

References

1. Ali, F., Suveatwatanakul, C., Nanu, L., Ali, M., Terrah, A.: Social media marketing and brand loyalty: exploring interrelationships through symmetrical and asymmetrical modeling. Spanish J. Mark. ESIC (2024). https://doi.org/10.1108/sjme-08-2023-0219
2. Aliu, J., Oke, A.E., Oluwasogo Oyediran, A., Kanya, R.A., Ukaha Onyeukwu, S.: From isolation to interaction: evaluating social media awareness and adoption in a developing construction industry. Smart Sustain. Built Environ. (2024). https://doi.org/10.1108/sasbe-01-2024-0027
3. Aljuhmani, H.Y., Elrehail, H., Bayram, P., Samarah, T.: Linking social media marketing efforts with customer brand engagement in driving brand loyalty. Asia Pac. J. Mark. Logist. **35**(7), 1719–1738 (2022). https://doi.org/10.1108/apjml-08-2021-0627
4. Anantharaman, R., Prashar, S., Tata, S.V.: Examining the influence of customer-brand relationship constructs and bandwagon effect on brand loyalty. Benchmark. Int. J. **30**(2), 361–381 (2022). https://doi.org/10.1108/bij-06-2021-0365
5. Beck, B., Moore Koskie, M., Locander, W.: How electronic word of mouth (ewom) shapes consumer social media shopping. J. Consum. Mark. **40**(7), 1002–1016 (2023). https://doi.org/10.1108/jcm-01-2023-5817
6. Cheung, M.L., Leung, W.K.S., Yang, M.X., Koay, K.Y., Chang, M.K.: Exploring the nexus of social media influencers and consumer brand engagement. Asia Pac. J. Mark. Logist. **34**(10), 2370–2385 (2022). https://doi.org/10.1108/apjml-07-2021-0522
7. Doğan, B., Balcioglu, Y.S., Elçi, M.: Multidimensional sentiment analysis method on social media data: comparison of emotions during and after the covid-19 pandemic. Kybernetes. https://doi.org/10.1108/k-09-2023-1808
8. Erhan, T.P., van Doorn, S., Japutra, A., Ekaputra, I.A.: Digital marketing innovation and firm performance: the role of decision-making comprehensiveness in dynamic environments. Asia Pac. J. Mark. Logist. **36**(2), 435–456 (2023). https://doi.org/10.1108/apjml-01-2023-0097

9. Hazzam, J.: The moderating role of age on social media marketing activities and customer brand engagement on Instagram Social Network. Young Cons. **23**(2), 197–212 (2021). https://doi.org/10.1108/yc-03-2021-1296
10. Ho-Mai, N.T., Tran, V.T., Nguyen, V.K., Do, U.T., Truong, T.B., Tran, P.T.: The pathways of increasing online celebrity brand equity and followers' hotel booking intention: a serial multiple mediation model. J. Res. Interact. Mark. (2024). https://doi.org/10.1108/jrim-08-2023-0265
11. Itani, O.S., Badrinarayanan, V., Rangarajan, D.: The impact of business-to-business salespeople's social media use on value co-creation and cross/up-selling: the role of social capital. Eur. J. Mark. **57**(3), 683–717 (2022). https://doi.org/10.1108/ejm-11-2021-0916
12. Izogo, E.E., Mpinganjira, M.: Somewhat pushy but effective: the role of value-laden social media digital content marketing (VSM-DCM) for search and experience products. J. Res. Interact. Mark. **16**(3), 365–383 (2021). https://doi.org/10.1108/jrim-05-2021-0146
13. Kumar, S., Hsieh, J.-K.: How social media marketing Activities Affect brand loyalty? Mediating role of brand experience. Asia Pac. J. Mark. Logist. (2024). https://doi.org/10.1108/apjml-09-2023-0900
14. Li, H., See-To, E.W.K.: Source credibility plays the central route: an elaboration likelihood model exploration in social media environment with demographic profile analysis. J. Electron. Bus. Digit. Econ. **3**(1), 36–60 (2023). https://doi.org/10.1108/jebde-10-2022-0038
15. Parris, D.L., Guzmán, F.: Evolving brand boundaries and expectations: looking back on brand equity, brand loyalty, and brand image research to move forward. J. Prod. Brand Manage. **32**(2), 191–231 (2022). https://doi.org/10.1108/jpbm-06-2021-3528
16. Romero-Jara, E., Solanellas, F., López-Carril, S., Kolyperas, D., Anagnostopoulos, C.: The more we post, the better? A comparative analysis of fan engagement on social media profiles of Football Leagues. Int. J. Sports Mark. Spons. **25**(3), 578–599 (2024). https://doi.org/10.1108/ijsms-12-2023-0252
17. Sait, M.A., Ali, M.A., Almunawar, M.N., Haji Masri, H.M.: Understanding factors to digital wallet discontinuance intention among past users: an exploratory study. J. Sci. Technol. Policy Manage. (2024). https://doi.org/10.1108/jstpm-01-2024-0005
18. Sakas, D.P., Giannakopoulos, N.T., Terzi, M.C., Kamperos, I.D., Kanellos, N.: What is the connection between fintechs' video marketing and their vulnerable customers' brand engagement during crises? Int. J. Bank Mark. (2023). https://doi.org/10.1108/ijbm-03-2023-0142
19. Senanu, B., Anning-Dorson, T., Tackie, N.N.: Social media insights for non-luxury fashion smes in emerging markets: evidence from young consumers. J. Fash. Mark. Manage. Int. J. **27**(6), 965–987 (2023). https://doi.org/10.1108/jfmm-02-2022-0026
20. Shabbirhusain, R.V., Annamalai, B., Chandrasekaran, S.: Global multi-sport events: content strategy for driving fan engagement on Twitter. Sport, Bus. Manage. Int. J. **13**(4), 450–469 (2023). https://doi.org/10.1108/sbm-10-2022-0094
21. Shahidzadeh, M.H., Shokouhyar, S.: Unveiling just-in-time decision support system using social media analytics: a case study on reverse logistics resource recycling. Ind. Manag. Data Syst. **124**(6), 2251–2283 (2024). https://doi.org/10.1108/imds-12-2023-0921
22. Silva, S.C., Duarte, P.A., Almeida, S.R.: How companies evaluate the ROI of social media marketing programmes: insights from B2B and B2C. J. Bus. Indus. Mark. **35**(12), 2097–2110 (2020). https://doi.org/10.1108/jbim-06-2019-0291
23. Tafesse, W., Wood, B.P.: Social Media influencers' community and content strategy and follower engagement behavior in the presence of competition: an Instagram-based investigation. J. Prod. Brand Manage. **32**(3), 406–419 (2022). https://doi.org/10.1108/jpbm-02-2022-3851
24. Venciute, D., Auruskeviciene, V., Reardon, J.: The impact of social media marketing on new venture performance. Corp. Commun. Int. J. **28**(5), 788–810 (2023). https://doi.org/10.1108/ccij-11-2022-0137
25. Villumsen, K., Elmer, H., Schmeltz, L.: Hosting and hoping on social media—a study on some communication strategies among Danish cultural institutions and tourist attractions during COVID-19. Corp. Commun. Int. J. **29**(5), 730–752 (2024). https://doi.org/10.1108/ccij-06-2023-0083

26. Vlachvei, A., Notta, O., Koronaki, E.: Effects of content characteristics on stages of customer engagement in social media: investigating European wine brands. J. Res. Interact. Mark. **16**(4), 615–632 (2021). https://doi.org/10.1108/jrim-12-2020-0275
27. Wei, L.H., Huat, O.C., Thurasamy, R.: The impact of social media communication on consumer-based brand equity and purchasing intent in a pandemic. Int. Mark. Rev. **40**(5), 1213–1244 (2023). https://doi.org/10.1108/imr-12-2021-0353
28. Zheng, X.: How does a firm's digital business strategy affect its innovation performance? an investigation based on knowledge-based dynamic capability. J. Knowl. Manage. (2024). https://doi.org/10.1108/jkm-05-2023-0410

Strategic Marketing Analytics: A Systematic Review

Ananya Mishra and Pradeep Kautish

Abstract Since its inception, strategic marketing analytics has attracted substantial scholarly attention and has continuously spread across a number of research areas. The training programs, assessment techniques, and theoretical underpinnings of strategic marketing analytics have all undergone constant evolution due to research. A thorough literature evaluation is still missing, though. Using bibliometric analysis, this study provides a comprehensive retrospective on strategic marketing analytics research, charting its development. Making use of bibliographic coupling, co-citation, and citation approaches, the study examines research trends by analysing 61 articles from the SCOPUS database (2000–2024) in order to identify intellectual and thematic frameworks. Three fundamental and seven theme clusters are identified by the study using VOSviewer for network analysis and visualisation. By highlighting the current state of strategic marketing analytics research, the findings broaden the field's scope.

Keyword Strategic marketing analytics · Marketing research evolution · Bibliometric analysis · Research trends

1 Introduction

In order to identify industry patterns, comprehend market phenomena, enhance the success of marketing initiatives and provide information for marketing decision-making, marketing professionals must periodically collect, manage, analyse, and report on data in environments rich in data. To do this, they need to employ tools and technologies that can quickly and effectively combine consolidate big data and produce insightful information that is relevant to all touchpoints. This is why

---

A. Mishra (✉) · P. Kautish
Department of Marketing, Institute of Management, Nirma University, Ahmedabad, Gujarat, India
e-mail: 223103@nirmauni.ac.in

P. Kautish
e-mail: pradeep.kautish@nirmauni.ac.in

S. Kautish et al. (eds.), *Strategy Analytics for Business Resilience Theories and Practices*, Information Systems Engineering and Management 33,
https://doi.org/10.1007/978-3-031-82369-5_12

marketing analytics is an essential component of marketing. A thorough summary of the "what," "where," "who," "when," "how," and "so what" of marketing analytics is provided by this original definition [6]. Applications for marketing analytics fall into two categories: those that integrate user input into one that uses an automated marketing analytics platform and others that do not. Technology-enabled and model-supported, marketing analytics is a method that "harnesses customer related marketing information driven data to augment decision making for consumer insights or related factors" [14].

Academics and industry professionals in the discipline of several backgrounds from which marketing analytics originate, including operations research, expert systems, marketing, and statistics [8]. In today's dynamic business environment, the capacity to leverage data-driven insights has become essential for companies seeking to stay ahead of the competition. In order to improve decision-making and boost marketing effectiveness, strategic marketing analytics includes the methodical gathering, examination, and interpretation of marketing data [33]. Robust analytical frameworks are more important than ever when businesses are flooded with information gathered from multiple sources, including social networking sites, consumer interactions, and market trends [6]. Through the application of marketing analytics, companies may evaluate the effectiveness of their marketing campaigns, predict future trends, allocate resources more efficiently, and modify their approaches to accommodate changing customer demands. Businesses can gain practical insights from this strategic strategy that can guide their pricing, product development, and marketing efforts [13]. Furthermore, businesses are depending more and more on cutting-edge analytical methods like consumer segmentation, machine learning, and predictive modelling to improve their marketing strategies as a result of the growth of digital marketing channels and the abundance of real-time data [20].

Prior analyses of strategic marketing analytics have provided a variety of incomplete insights from both qualitative and quantitative methods in their respective fields; nevertheless, a thorough summary of the body of research in the field is still missing. Additionally, there is no empirical evidence to support the particular drivers of the literature's growth in this quickly developing topic. By identifying the subjects that have attracted a lot of attention, new trends, and areas that require more research, bibliometric analysis can greatly improve our understanding of strategic marketing analytics research. This strategy might support the use of evidence-based approaches in a variety of disciplines. In order to gain deeper understanding in this field, the following research questions require more investigation:

1. What is the strategic marketing analytics research pattern?
2. What are the influential sources in strategic marketing analytics research?
3. In this field, which articles are the most cited and influential?
4. What are the main tenets of strategic marketing analytics research nowadays, as well as its practical applications?

We seek to provide a thorough and thorough examination of the body of knowledge on strategic marketing analytics research in order to respond to these research issues. We do this by using bibliometric analysis, which consists of a variety of quantitative methods able to work with sizable literature-related datasets [10, 25]. We examine and review the corpus of data that is currently available using a combination of science mapping and performance analysis. By using citation analysis, we may look at information from reputable writers, organisations, sources, and publications. A framework for further research in the wide domain of strategic marketing analytics is provided by co-citation based analysis and bibliographic coupling, which also assist in identifying underlying themes and fundamental information in the research on strategic marketing analytics.

To contribute significantly to the body of knowledge on strategic marketing analytics.

This study, which traces the evolution of the idea of strategic marketing analytics using bibliometric analysis, is one of the first to offer a thorough evaluation of the field. To gain valuable insights into core research areas, bibliographic networks, unified study clusters, and esteemed publication locations, we analyse 61 publications published over the course of more than 20 years. As a result, this study provides a comprehensive assessment of strategic marketing analytics research that has never been done before in the industry. Our analysis offers a thorough summary of the state of strategic marketing analytics research today by identifying three fundamental clusters and seven topical clusters. These insights add to the originality of this work in addition to directing future research areas.

2 Theoretical Framework of Strategic Marketing Analytics

2.1 Review Research Using Bibliometrics Analysis

With an average of 1021 yearly publications in the last ten years, the use of bibliometric generated analysis has become increasingly widespread in the social sciences discipline including economics, marketing, management, and business since the 1950s and has had a significant impact on exploration [10]. Two things are responsible for the growth: first, the emergence of scientific databases like Scopus and the Web of Science (WOS), which made it easier to access large bibliometric datasets; second, the creation of practical data analysis tools such as R Studio, VOSviewer, and Gephi, which stimulated bibliometric review of academic interest [10]. The present research attempts to provide a bibliometric summary of the broad field of strategic marketing analytics research through the use of scientific mapping methodologies and performance analyses.

3 Method

We employed quantitative methods and a bibliometric methodology [27]. We chose this method over systematic literature review (SLR) since it thoroughly analyses data to provide insights in multiple domains. The bibliometric analysis we conducted allowed us to visually represent important findings through the use of the VOS viewer application [31]. Nodes in the visualizations stand in for different components of the built network (including authors, references, and documents), and edges show how strongly these nodes are associated.

3.1 Selecting the Analysis Methods

Utilising many techniques for bibliometric analysis, such as co-citation analysis, citation analysis, performance analysis, and bibliographic coupling, we examined the structure of research on strategic marketing analytics, taking ideas from the suggestions made by [10]. Performance analysis offered a succinct overview of publication patterns linking strategic marketing analytics over time, and co-citation analysis of cited references was useful for assessing and describing the discipline's body of knowledge [10]. By concentrating on highly cited and influential works, co-citation analysis made it easier to identify foundational themes. Meanwhile, the bibliographic coupling of article clusters made it easier to understand the present developing themes in the field of study [4].

3.2 Data Collection

The SCOPUS provided the sample for the investigation. Because SCOPUS provides an extensive range of metadata essential for these kinds of assessments—such as references, author lists, citation counts, affiliations, geographic distribution, abstracts, and journal impact factors—we considered it relevant for bibliometric analysis [7]. Without regard to time constraints, every search was completed in a single day. We searched for "Strategic marketing analytics" in SCOPUS Topic Search and found 339 papers between 2000 and 2024.

For two main reasons, we only considered journal articles and did not include books, book chapters, or conference proceedings. First off, journal publications are released more regularly than books or book chapters, giving us access to recent data—which is crucial for keeping tabs on the most recent advancements and trends in the subject. Secondly, journal papers typically focus on certain research issues or hypotheses, offering specialised, condensed data that is especially well-suited for bibliometric analysis. This stage focused only on journal that was published papers in the English language, reducing the corpus size from 339 to 146 items.

Based on earlier research, we next narrowed the topic to concentrate only on studies pertaining to strategic marketing analytics. This covered disciplines including Business management, and accounting. This scope reduction resulted in the production of 94 articles.

And after the selection of subject area, keyword selection was so as to get the articles in specific area. This included keywords as Marketing Analytics, Marketing Strategy, Social Media Analytics, Customer Engagement, Customer Analytics, Social Media, Consumption Behavior, Social Media Strategy, Social Media Marketing, Consumer Attitude, Consumer Data Analysis, Corporate Strategy, Customer Service, Customer Segmentation, Customer Satisfaction, Customer Relationship Management, Green Marketing, Sustainable Development, Business Sustainability. 68 articles were produced as a result of this scope reduction.

Both researchers verified the data that was downloaded on their own. A manual screening process was used to choose 59 articles for inclusion from among the full-texts (or abstracts in the case that full-texts were not accessible) of the documents. Table 1 provides a summary of the inclusion criteria and comprehensive data retrieval procedures used in this investigation.

Table 1 Selection of documents and search parameters

Criteria for inclusion and exclusion and search	Exclusion	Inclusion
Search criteria		
Search engine: SCOPUS		
Keywords: "Strategic marketing analytics"		339
Source type: "Journal"	169	170
Document type: "article"	15	155
Publication stage: "final"	6	149
Language: "English"	3	146
Subject area: "Business, management and accounting"	52	94
Keyword selection: Marketing analytics, marketing strategy, social media analytics, social media, customer engagement, customer analytics, consumption behavior, social media strategy, social media marketing, consumer attitude, consumer data analysis, corporate strategy, customer service, customer segmentation, customer satisfaction, customer relationship management, green marketing, sustainable development, business sustainability	26	68
Article with no abstract	1	67
Article with no DOI	5	62
Duplicate documents	1	61

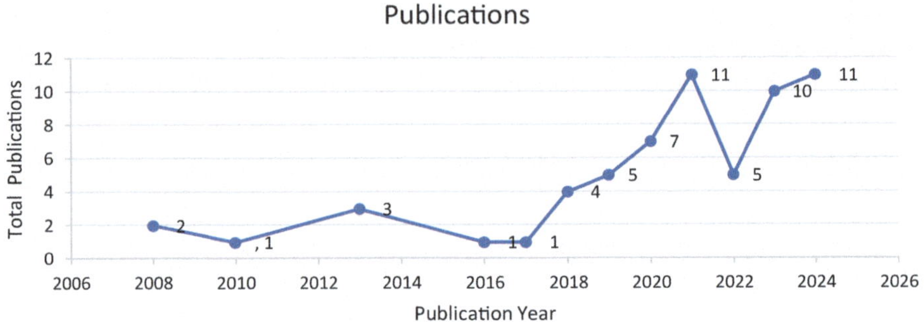

Fig. 1 Trends in strategic marketing analytics research publications

4 Outcome

4.1 Analytical Performance Evaluation of Strategic Marketing Research

4.1.1 Publication of Study on Strategic Marketing Analytics Research

Figure 1 shows the pattern of publications for strategic marketing analytics research between 2008 and 2024 by charting the published papers against the corresponding publication years. From 2018 onwards, we observe an optimistic shift in the pattern. With the most publications 2021 and 2024 are the most productive years, followed by 2020 with seven publications and 2023 with ten. Following in line with an equivalent number of publications, or five, are the years 2022 and 2019. Since there is only data used in the analysis through 2024, we observe a declining tendency in the years 2017, 2016, 2010, and with only 1 publication. Given we expect an increase in publications following the rise in the average number of publications per year starting in 2018.

4.1.2 Leading Sources of Documents of Strategic Marketing Analytics Research

Table 2 presents a list of the top 5 sites for research on strategic marketing analytics. The discipline has significantly improved, according to the results, as a result of its publication in several interdisciplinary publications. It's interesting to note that all strategic marketing analytics research journals are featured in the Australian Business Deans Council (ABDC) 2022 Journal Ranking (updated March 15, 2023), with 3 journals listed in the 'A*' category and 2 in the 'A' category. With the most citations 239, the leading source is the journal Industrial Marketing Management (ABDC "A*" category). It is followed by Technological Forecasting & Social Change (ABDC "A" category) publication with 209 citations. European Journal of Marketing (ABDC

"A*" category) publication with 79 citations. International Journal of Information Management (ABDC "A*" category) publication with 27 citations and Industrial Management and Data Systems (ABDC "A*" category) journal with 9 citations.

By the total number of publications, Technological Forecasting and Social Change tops the list, followed by four publications by Industrial Marketing Management with 3 publications.

Even though other list journals have less articles per journal, there is a growing trend in robustness across a range of study directions in the distribution and publishing efficiency. This implies that prestigious publications are open to publishing strategic marketing analytics research. Furthermore, it is clear from a closer look at the publishing trends in these highly cited journals from 2000 to 2024 that there will be an exponential growth in the quantity of publications in the area of strategic marketing analytics research between 2018 and 2024. Scholarly interest is sparked by publications in prestigious journals [5, 26]. This discernible surge in research activity on strategic marketing analytics points to an upward trajectory starting in 2024.

To evaluate the journals' influence, Table 3 also includes the Cite Scores of the Top 5 Journals. International Journal of Information Management, Technological Forecasting and Social Change and Industrial Marketing Management have the possess the highest Cite Scores 53.1, 21.3, and 17.3 citations, respectively among the Top 5 Journals. A prestige metric called SCImago Journal Rank (SJR) is predicated notion which "citations are not all made equal." According to SJR, the citation value is directly impacted by the topic field, journal quality, and reputation in addition to the overall quantity of citations in the subject related domains. The top three distinguished journals, as measured by SJR above the others on the list, are International Journal of Information Management, Technological Forecasting and social change, and Industrial Marketing Management with 5.775, 3.118 and 2.705 respectively. Source-Normalized Impact per Paper (SNIP) assesses references according to total number of citations in a subject area to ascertain a source's contextual citation efficacy. It makes it possible for researchers to compare sources directly on a variety of subjects. With SNIP ratings of 5.825, 2.945, and 2.167, respectively, the top three journals are International Journal of Information Management, Technological Forecasting and Social Change, and Industrial Marketing Management. As a result, the International Journal of Information Management has the greatest Cite score, SNIP, and SJR.

4.1.3 Top Articles on Strategic Marketing Analytics Research

The most top articles on strategic marketing analytics are included in Table 3. Among the top 10 listed articles, Sivarajah et al. [29] seminal work titled as "Role of big data and social media analytics for business-to-business sustainability: A participatory web context" has received most citations 171. This study has used exploratory qualitative study which closes the gap by utilizing information gleaned from manager interviews a as well as by creating taxonomies that emphasise the advantages of

Table 2 Prominent sources of strategic marketing analytics research

Sources (journals)	TC	ABDC	SJR (2023)	SNIP (2023)	Cite score Scopus (2023)	TP	Number of papers published every 5 years				
							2000–2005	2006–2011	2012–2017	2018–2023	2024
Industrial marketing management	239	A*	2.705	2.167	17.3	3				3	
Technological forecasting and social change		A	3.118	2.945	21.3	4				4	
European journal of marketing		A*	1.266	1.246	6.9	2			1	1	
International journal of information management		A*	5.775	5.825	53.1	2		1		1	
Industrial management and data systems		A	209	1.306	9.6	2				1	1
			79								
			27								
			9								

Table 3 Most influential articles on strategic marketing analytics research

Author	Title	Total citation
Sivarajah et al. [29]	Role of Big Data and Social Media Analytics for Business-to-Business Sustainability: A Participatory Web Context	171
Kitchens et al. [24]	Advanced Customer Analytics: Strategic Value Through Integration of Relationship-Oriented Big Data	170
Fildes et al. [12]	Retail forecasting: Research and practice	113
Wright et al. [36]	Adoption of Big Data Technology for Innovation in B2B Marketing	88
Huang and Rust [19]	IT-Related Service: A Multidisciplinary Perspective	87
Gupta et al. [15]	Big data and Firm Marketing Performance: Findings from a knowledge-based view	84
Tafesse and Wien [30]	Implementing social media marketing strategically: an empirical assessment	78
Aksoy [1]	How do you measure what you can't define?: The current state of loyalty measurement and management	75
Eriksson et al. [11]	Think with me, or think for me? On the future role of artificial intelligence in marketing strategy formulation	69
Quinn et al. [28]	Troubled waters: the transformation of marketing in a digital world	63

the participatory web in comparison to the passive online for conducting business operations for various kinds of organisations.

Kitchens et al. [24] carried out the second seminal paper titled as "Advanced Customer Analytics: Strategic Value Through Integration of Relationship-Oriented Big Data," which received 170 citations. The framework for locating and assessing diverse big data sources is developed in this article with the goal of building a data infrastructure with value justification that facilitates the targeted and efficient implementation of sophisticated customer analytics. To further leverage these rich constructs, create a customised kernel-based learning technique. Finally, implement demonstrating the framework's capabilities in a cutting-edge prototype system that can reliably forecast a range of consumer actions in a demanding setting substantial value-adding potential.

Another research titled as "Retail Forecasting: Research and Practice" by [12] is the third most influential study on the list, with 113 citations. The author's goal was to review the body of knowledge regarding retail demand forecasting. It explores how demand projections are used by retailers to support decision-making processes related to store location, product restocking, and promotions at different levels—strategic, tactical, and operational. It emphasizes the need for precise retail forecasting to enhance decision-making throughout supply chains. It also identifies several areas in which significant advancements could be made by future study. According to the data, causal models perform better in retail forecasting than simple benchmarks, but machine-learning approaches have not yet shown appreciable increases in accuracy. Forecasting models are made more complex by the effects of promotions, holidays, and social media data; yet, the impact of new variables such as online product reviews is still not well studied. It also points out that merchants frequently use simple time-series models for product forecasting, even though more sophisticated methods have the potential to improve the process. Forecasting new products is still very difficult because of the dearth of historical data.

4.2 Strategic Marketing Analytics Research's Intellectual and Influential Framework

4.2.1 Knowledge Foundation of Strategic Marketing Analytics Research: Co-citation Analysis

We were able to pinpoint the foundational publications and ideas that have influenced the area of strategic marketing analytics research by co-citation analysis of the literature that was cited. The co-citation map of references that the review corpus's articles cited at least once is shown in Fig. 2. The co-citation patterns of ten items were analyzed using VOSviewer with the full counting method, and the results showed that there are three main foundational clusters of research on strategic marketing analytics: the largest cluster (red nodes) is the critical keyword based analysis of big data challenges; the second largest cluster (green nodes) is the assessment of the business value of big data analytics; and the third cluster (blue nodes) is big data related basic concepts, methods, and analytics in social media and marketing analytics.

4.2.2 Strategic Marketing Analytics Research Theme Clusters: Bibliographic Coupling

Bibliographic coupling allows us to identify previously unrecognized contributions and new developing trends in the field of strategic marketing analytics by concentrating on publications that may not have garnered many citations. This method provided a more nuanced view of the changing domain of strategic marketing analytics research and assisted in capturing a wider range of literature [10].

The underlying principle of it is that publications having similar references also contain similar information [23, 35]. This purposeful citation of articles marks important information about the dominant topic structure of strategic marketing analytics by citing important publications in the field of strategic marketing analytics. The most significant and linked components within the network are highlighted in this analysis, which provides insightful information about the fundamental ideas and broad

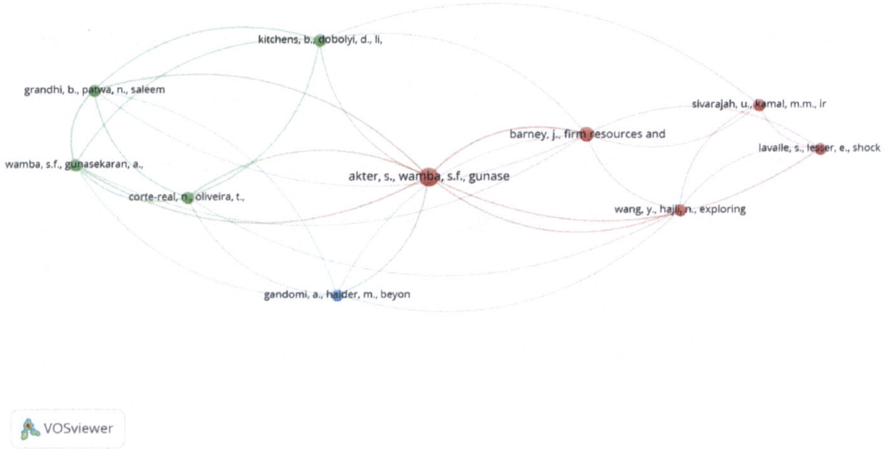

Fig. 2 Co-citation about strategic marketing analytics

patterns influencing the field of study [32]. Figure 3 depicts the seven various clusters via a network of bibliographic couplings, namely, *Big Data and Artificial Intelligence Publications in Marketing Strategy Analytics* in red nodes, *Leveraging Big Data and Predictive Analytics for Sustainable Supermarket Practice* in green nodes, *Transforming Manufacturing and Marketing Through Generative AI and Digital Technologies* in blue nodes, *Transforming Retail and B2B Marketing: The Role of Data Analytics and Social Media in Enhancing Customer Experience and Organizational Capabilities* in yellow nodes, *Bridging the Gap: Enhancing Analytical Skills in Marketing Education Through Social Listening* in purple nodes, *Enhancing Value Creation and Mitigating Business Failure through Customer Analytics and Marketing Analytics* in light blue, and *Enhancing Customer Engagement through Analytics and Click Tracking Strategies* in orange nodes. In this case, the nodes stand for publications, while the edges show the frequency with which two publications have cited a third article. Two publications have a stronger bibliographic coupling when there are more shared references between them. The SCOPUS database's 7 subject clusters, which support the current knowledge structure of strategic marketing analytics research, are succinctly summarised in Table 4.

Cluster 1 (largest cluster in red nodes) includes 12 *Big Data and Artificial Intelligence Publications in Marketing Strategy Analytics* that have received 525 citations. With 170 and 75 citations, respectively, Kitchens et al. [24] and Aksoy [1] are two of this cluster's most referenced articles. Kitchens et al. [24] articles carried out the methodology for finding and assessing different big data sources and put into practice the goal of building a value-based data infrastructure which permits the targeted and efficient use of sophisticated customer analytics. To make better use of these rich components, create a customized kernel-based learning approach. Finally, demonstrate the framework's tremendous value-adding potential by putting it into practice

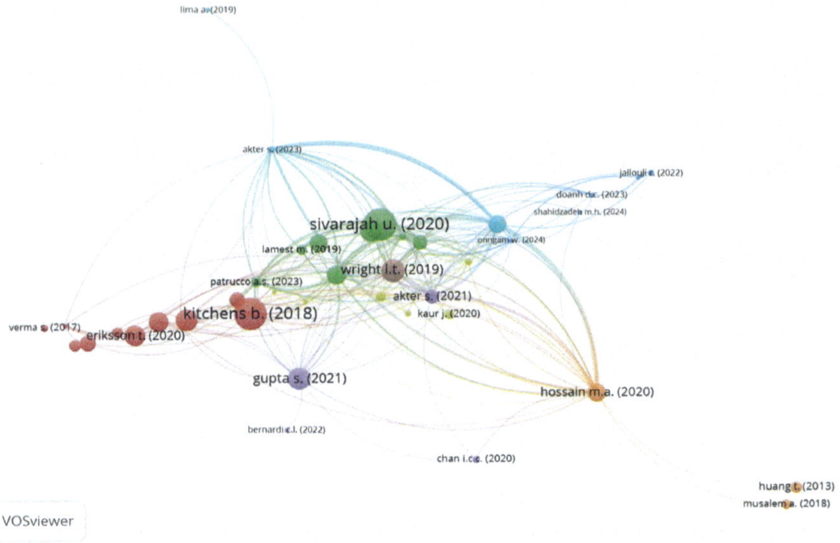

Fig. 3 Bibliographic coupling of articles on strategic marketing analytics research

in an inventive system prototype that successfully predicts a range of customer behaviors in a demanding context. On other hand, the study by Aksoy [1] investigates the discrepancy in customer loyalty measurement and management between scientific advice and practical implementations. According to the study, which is based on interviews with top marketing managers, most businesses value client loyalty, but not all of them have a defined definition. Metrics that are often monitored, such as customer happiness and recommendation likelihood, are not very good at forecasting loyalty-related actions. Furthermore, a lot of businesses employ rudimentary analytics, and just half look at how loyalty affects financial results. In the report, companies are urged to use advanced analytics, adopt more precise definitions, and more effectively connect loyalty indicators to financial performance.

Cluster 2 (green nodes) includes 8 articles on *Leveraging Big Data and Predictive Analytics for Sustainable Supermarket Practice* which have been cited 438 times in Scopus. In this cluster, the two most cited In this cluster, the two most referenced articles are articles are Sivarajah et al. [29] and Tafesse and Wien [30] with 171 and 78 citations. Sivarajah et al. [29] conducted qualitative research on an issue that has received little attention in the business and management literature: the participatory web's role in facilitating sustainable business-to-business (B2B) activities. It illustrates how businesses may increase productivity, efficiency, and sustainability in their operations and marketing departments by utilizing participatory online technologies like big data and social media analytics. The research creates taxonomies from manager interviews that demonstrate how the interactive web outperforms passive online environments in supporting strategic business activities. In the study of Tafesse and Wien [30], they talk about how firms use social media

Table 4 Thematic clusters of strategic marketing analytics research

Theme	Author(s)	Title (the top two articles mentioned for each theme cluster)	TC
AI & Big Data in Marketing Strategy Analytics	Kitchens et al. [24]	Advanced Customer Analytics: Strategic Value Through Integration of Relationship-Oriented Big Data	170
	Aksoy [1]	How do you measure what you can't define?: The current state of loyalty measurement and management	75
Leveraging Big Data and Predictive Analytics for Sustainable Supermarket Practices	Sivarajah et al. [29]	Role of Big Data and Social Media analytics for business to Business Sustainability: A Participatory Web Context	171
	Tafesse and Wien [30]	Implementing social media marketing strategically: an empirical assessment	78
Transforming manufacturing and Marketing Through Generative AI and Digital Technologies	Jallouli and Kaabi [21]	Mapping Top Strategic E-commerce Technologies in the Digital Marketing Literature	8
	Doanh et al. [9]	Generative AI in the manufacturing process: theoretical considerations	5
Transforming Retail and B2B Marketing: The Role of Data Analytics and Social Media in Enhancing Customer Experience and Organizational Capabilities	[22]	Influence of technological advances and change in marketing strategies using analytics in the retail industry	18
	Wang et al. [34]	Exploiting business networks in the age of social media: the use and integration of social media analytics in B2B marketing	17
Bridging the Gap: Enhancing Analytical Skills in Marketing Education Through Social Listening	Gupta et al. [15]	Big data and firm marketing performance: Findings from knowledge-based view	84
	Akter et al. [2]	Big data-driven strategic orientation in international marketing	37
Enhancing Value Creation and Mitigating Business Failure through Customer Analytics and Marketing Analytics	Hossain et al. [17]	Why doesn't our value creation payoff: Unpacking customer analytics-driven value creation capability to sustain competitive advantage	55
	Akter et al. [3]	How does quality-dominant logic ensure marketing analytics success and tackle business failure in industrial markets?	7

(continued)

Table 4 (continued)

Theme	Author(s)	Title (the top two articles mentioned for each theme cluster)	TC
Enhancing Customer Engagement through Analytics and Click Tracking Strategies	Hossain et al. [16]	Revisiting customer analytics capability for data-driven retailing	58
	Huang and Van Mieghem [18]	The promise of strategic customer behavior: On the value of click tracking	23

in a logical approach to assist their strategic marketing efforts. It sees social media deployment as a complex organizational architecture that includes social media presence, strategy, consumer interaction initiatives, and analytics. The study investigates the effects of these factors on company performance using primary data, making a distinction between marketing and social media performance. The findings demonstrate that all factors—aside from active presence—have a good impact on social media performance, which in turn has a beneficial impact on marketing performance. By presenting and empirically validating this novel social media implementation construct, the study adds to the body of literature.

Cluster 3 (blue nodes) includes 6 articles on *Transforming manufacturing and Marketing Through Generative AI and Digital Technologies* which is cited 14 times. The top 2 cited articles in this cluster are Jallouli and Kaabi [21] and Doanh et al. [9] has 8 and 5 citations respectively. Doanh et al. [9] study looks at how industrial processes are affected by digital transformation, specifically concerning Generative Artificial Intelligence (GAI). Product design, workforce optimization, demand forecasting, marketing strategy, quality assurance, and predictive maintenance are just a few of the Industry 4.0 aspects that are examined. It illustrates how GAI supports these areas through process optimization, product design enhancement, quality control enhancement, and innovation promotion using a qualitative, critical review approach. Predictive analytics for demand forecasting, quality problems, risk prediction, and market trend detection are also made possible by GAI, which aids managers in enhancing industrial efficiency and strategy. Jallouli and Kaabi [21] article offers an extensive management literature assessment of e-commerce technology. Eleven categories are used to classify e-commerce technology, and their effects on digital marketing tactics are examined. The review emphasizes how platforms and information and communication technologies (ICTs) are most commonly referenced in the literature, with a growing focus on artificial intelligence. It further demonstrates the focus that digital marketing research has placed on social media platforms, targeting techniques, and segmentation made possible by technology. The report does, however, highlight a deficiency in the literature about the effects of blockchain, IoT, cloud computing, and data analytics, and it recommends further research in these areas.

Cluster 4 (yellow nodes) includes 6 articles on *Transforming Retail and B2B Marketing: The Role of Data Analytics and Social Media in Enhancing Customer Experience and Organizational Capabilities* which is referenced 55 times. The top

2 cited articles are Kaur et al. [22] and Wang et al. [34] with 18 and 17 citations respectively. Kaur et al. [22] examines data analytics and technology developments are revolutionizing the retail sector, increasing consumer pleasure and decreasing attrition. It demonstrates how customer experience can be improved by fusing marketing ideas like market basket modeling, value-based consumer segmentation, and campaign planning with technological advancements like video, social media, wireless analytics, and smart vision systems. To give merchants a competitive edge in the retail industry, the paper seeks to explore how these technologies and analytics might help them attract and keep customers through the development of efficient merchandising and marketing strategies. Wang et al. [34] address the issue of organizational competency in business-to-business (B2B) marketing, he suggests a social media analytics development model (SMA). Optimising the advantages of commercial networks in the social media age and incorporating SMA into B2B marketing are the goals of the approach. Four interrelated layers of SMA capability are identified by the study: technological, operational, managerial, and strategic. This is done through a critical interpretive synthesis of academic, industry, and magazine publications. The technological prowess of SMA is combined with these levels to create dynamic organizational capabilities that facilitate skill development, performance evaluation, cultural shifts, and strategic planning. Technology serves as the cornerstone for wider organizational capacities, and the study emphasizes socio-technical phenomena in B2B marketing in SMA.

Cluster 5 (purple nodes) includes 5 articles on *Bridging the Gap: Enhancing Analytical Skills in Marketing Education Through Social Listening* that have been cited 135 times. The top cited articles are Gupta et al. [15] and Akter et al. [2] with 84 and 37 citations respectively. Akter et al. [2] provide a thorough framework for improving businesses' long-term success in foreign markets, this study thoroughly examines the literature on big international marketing strategies with a data-driven focus. Following the analysis of 2242 articles, 45 pertinent papers were combined to determine the three main orientations—the approach towards international markets, the orientation towards international innovation and entrepreneurship, and the orientation towards international digital platforms—each of which is backed by 11 subdimensions. According to the research, data-driven and data-evolving strategies are essential for successful international marketing because they allow businesses to take advantage of cutting-edge analytics and realize market value. This study fills in a major knowledge vacuum by offering managers a multifaceted framework for strategic direction in an international company. However, it also makes clear that more research is required to focus on certain firm types. Gupta et al. [15] investigate how big data analytics fits into Industry 4.0. The research highlights key elements for adopting big data in practical contexts by creating a conceptual model based on a knowledge-based view. The study demonstrates how big data improves strategic and operational decisions, which in turn leads to better marketing performance, and reveals important holes in conventional decision-making processes through the use of qualitative and thematic analysis informed by a knowledge-based theoretical framework. The suggested model offers managers a point of reference when trying to coordinate their company's operations for more successful marketing initiatives. It also offers a conceptual framework for further empirical research in this field.

Cluster 6 (light blue nodes) includes 3 articles on *Enhancing Value Creation and Mitigating Business Failure through Customer Analytics and Marketing Analytics* cited 65 times. The top 2 cited articles are Hossain et al. [17] and Akter et al. [3] with 55 and 7 citations respectively. As per Hossain et al. [17] value creation has been greatly improved by recent developments in data-driven marketing analytics, which has raised interest in this area among academics and business leaders. This study aims to close the knowledge gap on customer analytics-driven value creation capabilities (CAVCC) by concentrating on the unique role that customer analytics plays in value creation. For CAVCC, the study uses the resource-based view (RBV), knowledge-based view (KBV), and market orientation (MO) to create and evaluate a hierarchical model within the context of big data analytics. The results show that persistent competitive advantage is impacted both directly and indirectly and that customer linking plays a critical mediating role in this process. Akter et al. [3] study looks into how marketing analytics might lower the rate of business failure in industrial markets, especially in the post-pandemic environment, where there isn't much research yet. The study creates a model that determines the predictors of marketing analytics success (MAS) based on data, model, and deployment quality by utilizing a quality-dominant logic and dynamic capabilities framework. The results of a study done in Australia with 314 participants show that MAS considerably increases strategic company value and profitability. To address the reasons behind company failure, this research provides a theoretically sound and practically useful framework for utilizing analytics findings.

Cluster 7 (orange nodes) includes 3 articles on *Enhancing Customer Engagement through Analytics and click-Tracking Strategies* cited 81 times. The top 2 cited articles are Hossain et al. [16] and Huang and Van Mieghem [18] with 58 and 23 citations respectively. Hossain et al. [16] address the paucity of research on customer analytics capabilities in this setting and emphasize the strategic significance of customer analytics in the competitive retail environment. It presents a multidimensional consumer analytics capabilities model that identifies six core dimensions and twelve sub-dimensions connected to the resource-based view (RBV) of retailers through methodical literature research and thematic analysis. Value creation (offering and customization abilities), value management (data management and security capabilities), and value delivery (distribution and communication capabilities)) are the three main marketing concepts that unite these dimensions. Through the application of this paradigm, practitioners can improve customer equity and successfully engage customers, leading to improved business outcomes. Whereas, the study by Huang and Van Mieghem [18] looks into the expanding practice of click monitoring and how it affects strategic consumers who might be reluctant to visit websites if they know their clicks are being tracked. The study investigates how product availability influences consumers' willingness to voluntarily provide advance demand information using a newsvendor framework. The results show that there is a strong Nash equilibrium, meaning that all customers have an incentive to click and that their motivation doesn't change even in the presence of noise. This refutes the widely held notion that click tracking presents hazards for companies, emphasizing its advantages over tactics like advance selling and availability guarantees for both clients and companies. The study also takes into account how marketing and operational choices, such as markdown pricing and price-sensitive demand, interact to determine the total benefit of click tracking.

5 Conclusion

The SCOPUS database yielded 61 papers on strategic marketing analytics research (2000–2024) after exclusion–inclusion criteria were used. We conducted a systematic examination of publication patterns, notable sources, influential articles, knowledge foundations, and theme clusters using VOSviewer and bibliometric analysis. Since 2018, there has been an upward trend of publications, with an average of more publications per year. Leading sources show that top journals are open to strategic marketing analytics research: Industrial Marketing Management (ABDC "A*" category) has 239 citations, while Technological Forecasting & Social Change (ABDC "A" category) has 209 citations.

The journals with the greatest Cite Scores are the International Journal of Information Management, Technological Forecasting & Social Change, and Industrial Marketing Management, with 53.1, 21.3, and 17.3 citations, respectively. The journals Industrial Marketing Management (2.705), Technological Forecasting & Social Change (3.118), and International Journal of Information Management (5.775) have the highest SJR ratings. The International Journal of Information Management (5.825), Technological Forecasting & Social Change (2.945), and Industrial Marketing Management (2.167) are the top journals for SNIP values. With 171 citations, the top paper by Sivarajah et al. [29] is titled "Role of Big Data and Social Media Analytics for Business-to-Business Sustainability: A Participatory Web Context."

Co-citation analysis identified three key clusters: big data issues; big data concepts, methodologies, and analytics in social media and marketing analytics; and appraisal of the business value of big data analytics. The most significant and interrelated components in the network were highlighted in seven primary topic clusters found by bibliographic coupling, which were as follows: *Big Data and Artificial Intelligence Publications in Marketing Strategy Analytics, Leveraging Big Data and Predictive Analytics for Sustainable Supermarket Practice, Transforming manufacturing and Marketing Through Generative AI and Digital Technologies, Transforming Retail and B2B Marketing: The Role of Data Analytics and Social Media in Enhancing Customer Experience and Organizational Capabilities, Bridging the Gap: Enhancing Analytical Skills in Marketing Education Through Social Listening, Enhancing Value Creation and Mitigating Business Failure through Customer Analytics and Marketing Analytics*, and *Enhancing Customer Engagement through Analytics and Click Tracking Strategies*. Thus, through a thorough literature review, the study investigates recurrent themes and the condition of strategic marketing analytics research today. The results of this study offer a thorough overview of the significant intellectual framework in the subject of strategic marketing analytics, which has enormous research benefits for practitioners and scholars.

References

1. Aksoy, L.: How do you measure what you can't define? The current state of loyalty measurement and management. J. Serv. Manag. **24**(4), 356–381 (2013). https://doi.org/10.1108/josm-01-2013-0018
2. Akter, S., Hossain, M.A., Lu, Q., Shams, S.R.: Big data-driven strategic orientation in international marketing. Int. Mark. Rev. **38**(5), 927–947 (2021). https://doi.org/10.1108/imr-11-2020-0256
3. Akter, S., Hossain, M.A., Tarba, S.Y., Leonidou, E.: How does quality-dominant logic ensure marketing analytics success and tackle business failure in industrial markets? Ind. Mark. Manage. **109**, 44–57 (2023). https://doi.org/10.1016/j.indmarman.2022.12.005
4. Andersen, N.: Mapping the expatriate literature: a bibliometric review of the field from 1998 to 2017 and identification of current research fronts. Int. J. Human Resour. Manage. (2019). https://doi.org/10.1080/09585192.2019.1661267
5. Baker, H.K., Kumar, S., Pattnaik, D.: Research constituents, intellectual structure, and collaboration pattern in the Journal of Forecasting: a bibliometric analysis. J. Forecast. **40**(4), 577–602 (2021). https://doi.org/10.1002/for.2731
6. Basu, R., Lim, W.M., Kumar, A., Kumar, S.: Marketing analytics: the bridge between customer psychology and marketing decision-making. Psychol. Mark. **40**(12), 2588–2611 (2023). https://doi.org/10.1002/mar.21908
7. Birrer, D., Scalvedi, B., Frings, N.: A bibliometric analysis of mindfulness and acceptance research in sports from 1969 to 2021. Mindfulness **14**(5), 1038–1053 (2023). https://doi.org/10.1007/s12671-023-02124-5
8. Brito, P.Q., Soares, C., Almeida, S., Monte, A., Byvoet, M.: Customer segmentation in a large database of an online customized fashion business. Robot. Comp.-Integr. Manuf. **36**, 93–100 (2015). https://doi.org/10.1016/j.rcim.2014.12.014
9. Doanh, D.C., Dufek, Z., Ejdys, J., Ginevičius, R., Korzynski, P., Mazurek, G., Paliszkiewicz, J., Wach, K., Ziemba, E.: Generative AI in the manufacturing process: theoretical considerations. Eng. Manage. Produc. Serv. **15**(4), 76–89 (2023). https://doi.org/10.2478/emj-2023-0029
10. Donthu, N., Kumar, S., Mukherjee, D., Pandey, N., Lim, W.M.: How to conduct a bibliometric analysis: an overview and guidelines. J. Bus. Res. **133**, 285–296 (2021). https://doi.org/10.1016/j.jbusres.2021.04.070
11. Eriksson, T., Bigi, A., Bonera, M.: Think with me, or think for me? On the future role of artificial intelligence in marketing strategy formulation. TQM J. **32**(4), 795–814 (2020). https://doi.org/10.1108/tqm-12-2019-0303
12. Fildes, R., Ma, S., Kolassa, S.: Retail forecasting: research and practice. Int. J. Forecast. **38**(4), 1283–1318 (2022). https://doi.org/10.1016/j.ijforecast.2019.06.004
13. France, S.L., Ghose, S.: Marketing analytics: methods, practice, implementation, and links to other fields. Expert Syst. Appl. **119**, 456–475 (2018). https://doi.org/10.1016/j.eswa.2018.11.002
14. Germann, F., Lilien, G.L., Rangaswamy, A.: Performance implications of deploying marketing analytics. Int. J. Res. Mark. **30**(2), 114–128 (2012). https://doi.org/10.1016/j.ijresmar.2012.10.001
15. Gupta, S., Justy, T., Kamboj, S., Kumar, A., Kristoffersen, E.: Big data and firm marketing performance: findings from knowledge-based view. Technol. Forecast. Soc. Chang. **171**, 120986 (2021). https://doi.org/10.1016/j.techfore.2021.120986

16. Hossain, M.A., Akter, S., Yanamandram, V.: Revisiting customer analytics capability for data-driven retailing. J. Retail. Consum. Serv. **56**, 102187 (2020). https://doi.org/10.1016/j.jretconser.2020.102187
17. Hossain, M.A., Akter, S., Yanamandram, V.: Why doesn't our value creation payoff: unpacking customer analytics-driven value creation capability to sustain competitive advantage. J. Bus. Res. **131**, 287–296 (2021). https://doi.org/10.1016/j.jbusres.2021.03.063
18. Huang, T., Van Mieghem, J.A.: The promise of strategic customer behavior: on the value of click tracking. Prod. Oper. Manag. **22**(3), 489–502 (2013). https://doi.org/10.1111/j.1937-5956.2012.01386.x
19. Huang, M.-H., Rust, R.T.: IT-related service: a multidisciplinary perspective. J. Serv. Res. **16**(3), 251–258 (2013). https://doi.org/10.1177/1094670513481853
20. Iacobucci, D., Petrescu, M., Krishen, A., Bendixen, M.: The state of marketing analytics in research and practice. J. Mark. Anal. **7**(3), 152–181 (2019). https://doi.org/10.1057/s41270-019-00059-2
21. Jallouli, R., Kaabi, S.: Mapping top strategic e-commerce technologies in the digital marketing literature. J. Telecommun. Digit. Econ. **10**(3), 149–164 (2022). https://doi.org/10.18080/jtde.v10n3.554
22. Kaur, J., Arora, V., Bali, S.: Influence of technological advances and change in marketing strategies using analytics in retail industry. Int. J. Syst. Assur. Eng. Manag. **11**(5), 953–961 (2020). https://doi.org/10.1007/s13198-020-01023-5
23. Kessler, M.M.: Bibliographic coupling between scientific papers. Am. Doc. **14**(1), 10–25 (1963)
24. Kitchens, B., Dobolyi, D., Li, J., Abbasi, A.: Advanced customer analytics: strategic value through integration of relationship-oriented Big Data. J. Manag. Inf. Syst. **35**(2), 540–574 (2018). https://doi.org/10.1080/07421222.2018.1451957
25. Mishra, M., Mund, P.: Fifty-two years of consumer research based on social exchange theory: a review and research agenda using topic modeling. Int. J. Cons. Stud. **48**(4) (2024). https://doi.org/10.1111/ijcs.13074
26. Pattnaik, D., Hassan, M.K., Kumar, S., Paul, J.: Trade credit research before and after the global financial crisis of 2008—a bibliometric overview. Res. Int. Bus. Fin. **54**, Article 101287 (2020). https://doi.org/10.1016/j.ribaf.2020.101287
27. Pritchard, A.: Statistical bibliography or bibliometrics? J. Document. **25**(4), 348–349 (1969)
28. Quinn, L., Dibb, S., Simkin, L., Canhoto, A., Analogbei, M.: Troubled waters: the transformation of marketing in a digital world. Eur. J. Mark. **50**(12), 2103–2133 (2016). https://doi.org/10.1108/ejm-08-2015-0537
29. Sivarajah, U., Irani, Z., Gupta, S., Mahroof, K.: Role of big data and social media analytics for business to business sustainability: a participatory web context. Ind. Mark. Manage. **86**, 163–179 (2020). https://doi.org/10.1016/j.indmarman.2019.04.005
30. Tafesse, W., Wien, A.: Implementing social media marketing strategically: an empirical assessment. J. Mark. Manag. **34**(9–10), 732–749 (2018). https://doi.org/10.1080/0267257x.2018.1482365
31. Van Eck, N.J., Waltman, L.: Software survey: VOSviewer, a computer program for bibliometric mapping. Scientometrics **84**(2), 523–538 (2010)
32. Van Eck, N.J., Waltman, L.: Visualizing bibliometric networks. In: Measuring Scholarly Impact (2014). https://doi.org/10.1007/978-3-319-10377-8_13
33. Vollrath, M.D., Villegas, S.G.: Avoiding digital marketing analytics myopia: revisiting the customer decision journey as a strategic marketing framework. J. Mark. Anal. **10**(2), 106–113 (2021). https://doi.org/10.1057/s41270-020-00098-0
34. Wang, Y., Rod, M., Deng, Q., Ji, S.: Exploiting business networks in the age of social media: the use and integration of social media analytics in B2B marketing. J. Bus. Indus. Mark. **36**(12), 2139–2149 (2020). https://doi.org/10.1108/jbim-05-2019-0173
35. Weinberg, B.H.: Bibliographic coupling: a review. Inf. Storage Retrieval **10**(5–6), 189–196 (1974). https://doi.org/10.1016/0020-0271(74)90058-8

36. Wright, L.T., Robin, R., Stone, M., Aravopoulou, E.: Adoption of big data technology for innovation in B2B marketing. J. Business-to-Business Mark. **26**(3–4), 281–293 (2019). https://doi.org/10.1080/1051712x.2019.1611082

The manufacturer's authorised representative in the EU is Springer Nature Customer Service Centre GmbH, Europaplatz 3, 69115 Heidelberg, Germany. If you have any concerns regarding our products, please contact ProductSafety@springernature.com

Printed and bound by CPI Group (UK) Ltd, Croydon, CR0 4YY
01/10/2026
02225956-0005